ASTONISHING MESSAGES from the COSMOS

* * * * * *

UNBELIEVABLE, BUT TRUE STORY!

TAPED MESSAGES ARE PROOF!

HIGHLY EVOLVED SUPER BEINGS FROM FAR-OFF WORLDS APPEARED IN FULLY MATERIALIZED BODY FORMS. THEY EXPLAINED WHY UFO'S ARE SURROUNDING THE EARTH DURING ITS DRASTIC CHANGES. THEY PREDICT A MOST MARVELOUS NEW AGE FUTURE ON THE EARTH AFTER THE YEAR 2,000.

by LAURA HORST

* * * * * *

Library of Congress Catalog Card No.: 86-91286
ISBN: 0-910122-90-3

Printed in the United States of America by
Palmer Publications, Inc., Amherst, Wisconsin 54406

ACKNOWLEDGEMENT

Beginning January, 1957, the MASTER JESUS established a spiritual center in a modest, lovely home in Belmont Shores, Ca. There he could freely use Sarah Ferguson as a communicating channel for the transmission of messages from the spirit realms of life and the earth world. JESUS brought hundreds of lengthy explanatory messages covering varied topics.

That center became a central meeting place where a great number of very highly evolved GREAT SPIRIT MASTERS could congregate and bring their LOVE, POWER and unheard of, far advanced wisdom messages . . . for the benefit of all people.

There were about three hundred SPIRIT PERSONALITIES who brought messages. Some of their messages have been quoted in this book. However, there are hundreds of taped, transcribed messages that await future publications.

GRATEFUL THANKS to all SPIRIT PERSONALITIES who participated in this GREAT EVENT that took place in the home of Sarah (Sara).

JANUARY 1, 1957 . . . DECEMBER 8, 1966

CONTENTS

INTRODUCTION

THIS IS A TRUE STORY.

ASTONISHING? YES!

But for the skeptic, verification is available. For every word of it is backed by the recorded, dated, spoken words on tapes. The words as spoken were transmitted to tapes electronically, and immediately dated as to time and place. Later they were transcribed into typed copy. All records—tapes, typed pages and notes were carefully kept in consecutively dated steel storage containers. At the present time they fill a small studio.

The first tape recorded was dated January 14, 1957. The last transmission on tape was on December 9, 1966. The place was in a small home in Belmont Shores, an oceanside area of Long Beach, California.

Many who read this account will not be at all surprised that Highly Evolved Spirit Beings created and used the opportunity to give their informative messages to an unusually dedicated group of seven people.

Messages from other dimensions higher than the earth in evolution have been coming through to mankind all over the earth world for eons upon eons of time. But it is only lately, with our electronic devices, we have been able to record their words. The many different Holy Books now used by mortals contain similar counselings and prophecies.

However, none of the biblical and other accounts have given proof of the consistent many *materializations with tangible forms . . . of these highly developed Speakers, as in this book.* So tangible, in fact, that ordinary human beings not only heard and saw, but they were touched by those

Great Beings and conversed with them.

Every prophecy of note has foretold of the "great changes" that would take place upon the earth with the advent of the Aquarian Spiritual Age Cycle of Time. About one hundred years ago the earth entered this New Spiritual Age. All discerning persons must be sensing the accelerated spiritual awareness now being displayed by great masses of mortal people on the earth.

The passage of TIME means nothing to Spirit Beings dwelling in other dimensions of life. Events as they come about on the earth is their Time. For this book, that TIME is NOW!

* * * * *

The reader will want to know how it was that our group of seven people happened to come together in Belmont Shores. Unknown to us at that time, it was our Special Master Spirit Teachers who brought us together, because we were functioning on similar levels of our own accelerated spiritual awareness.

Our Master Spirit Teachers brought us to Sara's home. She was a lovely, most gracious psychic, greatly developed in clairvoyance and clairaudience. She had acquired the most unusual gift of "clear psychic seeing" with her spiritual sense of sight. There in Sara's home the numerous Highly Evolved Master Spirit Teachers established a "center." Each person in our group attended separate private sessions with Sara each week. During each session the Great Master Spirit Teachers used Sara as their vocal transmission channel to relay to us their fantastic informative messages.

During that long period of many years while the steady flow of highly advanced educational messages continued, there were about thirty months of that time when our Special Master Spirit Teachers fully materialized themselves and appeared before us. Also many other Great Spirit Beings materialized themselves.

Dear reader, we now are sharing with you the superb, priceless, astounding experiences enjoyed by each of us.

The members of our private group were:

Sara and her husband Ray, Catherine, Marian, Jean, my husband Tony and myself, Lara—as they called me.

All of these members have left the earthly dimension, having made their transition into spirit life in the spirit world. I, Lara, the scribe and the transcriber of this material, have been left here as the lone steward of these priceless records, to be used for present publication and for research.

Ever since these unusual events took place they have seemed so incredible—even to me, who saw and heard and witnessed all of them—I have waited until now to tell this startling story. Enough confirming evidence of the validity of these messages and their far-reaching predictions have now proven to be factual through actual worldwide happenings. Thus, the messages enable dependence to be placed upon the future predictions that are yet to take place on the earth BEFORE and AFTER the year 2,000. This book is the first of several proposed sequel series.

PREFACE

THE UNFOLDING UNIVERSE

Our ALMIGHTY CREATOR'S Vast Cosmic Plan of creation is ever unfolding. We now have entered into His "Glorious Plan of the Ages." It is the FABULOUS GOLDEN AGE . . . the AQUARIAN-KOSMON SPIRITUAL AGE . . . the GREATEST AGE THAT HAS EVER EXISTED ON THE EARTH.

To bring into manifestation and perfect operation, this present GOLDEN AGE CYCLE OF TIME, is a "Colossal Event." It requires the harmonious cooperation of numerous millions of Highly Evolved Spirit Beings, and hundreds of altruistic-minded earth people, all working together as strong links-in-a-chain.

Countless Illumined, Mighty, Powerful Beings, who, during thousands of years, have lived in the highly refined, far-off galaxies of worlds have now come to the earth, coming from their spirit realms of life to help bring this NEW AGE into operation.

They have acquired great wisdom and knowledge pertaining to the operation of many Vast Cosmic Schemes of Creations, having learned the universal laws. They have become highly skilled in controlling the universal elements, and the mighty cosmic electrical energies and powers that exist everywhere in the universal cosmos. They also have learned many Great Secrets of Life that are unknown to earth people.

The Mighty Celestial Master Beings and their millions of Co-Workers . . . their Righteous, Highly Evolved Spirit Hosts . . . are diligently working to cleanse the earth and its lower heavens, to expose and eliminate all the wrong, evil things that now are so rampant upon the earth. They are working to bring about righteousness on the earth and in the earth's lower heavens. This cleansing must first be done before this present NEW AGE can be firmly established.

THE SOLAR SYSTEM OPERATES IN ARCS OF LIGHT

The Great Master Teachers explain that the entire solar system containing the earth world operates within its own separate cyclic coil-phalanx of mighty electrical powers. During past cycles of time, our cyclic-coil has been traveling (like a great serpent), and rising higher and ever higher, as it has journeyed through many ARCS OF LIGHT. Each ARC has its own graded degree of Spiritual Light. It requires one cycle of about 3,000 years for our cyclic-coil to travel through one Arc of Light. It is estimated that about one hundred years ago, our cyclic-coil started its travel through the Greater, Higher, Brighter SPIRITUAL ARC OF KOSMON, located in the high etherean firmament that exists above our atmospherean firmament. Therefore, the entire earth world is undergoing very drastic changes of GREAT MAGNITUDE.

Those drastic earth changes are being brought about by earthquakes, tornados, floods, tidal waves, holocausts and catastrophes of various kinds. Thereby, they cause deaths of thousands of people. During those changes, some lands will disappear, and others will arise from the depths of the oceans, such as the ancient submerged continents of Atlantis and Lemuria. It is foretold those earth changes will continue through the year 2,000.

ALL THOSE AWESOME EARTH CHANGES ARE BRINGING ABOUT THE THOROUGH CLEANSING OF THE EARTH, cleansing if from all iniquities, unrighteous people on the earth and the evil entities in the lower heavens. This cleansing is preparing the earth for the time when this NEW SPIRITUAL GOLDEN AGE begins to function.

This thorough cleansing also includes all planets and

worlds in the earth's solar system. The cleansing affects all earth people and all Immortal Spirit beings living in the earth's heavenly spirit realms of life. This cleansing is changing ALL THINGS upon the earth, and ALL THINGS ARE BEING MADE NEW.

This NEW GOLDEN AGE will manifest many wonderful things, and many unusual things will come from "out of the air." And when the GREAT DAY arrives, then the heavens and the earth will be functioning together as "ONE PLACE." Then earth people and the so-called "dead" Mortals, who are our departed loved ones and friends, who now dwell in the earth's heavens, will be able to mingle together in joyous happy reunions.

AT THAT TIME, IN THAT ASTONISHING FUTURE, GLORIOUS, LOVE-FILLED AND WONDERFUL PEOPLE WILL PREDOMINATE THE EARTH AND THEY WILL LIVE IN PEACE AND HARMONY.

PART 1

UNBELIEVABLE, BUT TRUE

UNBELIEVABLE, BUT TRUE

As this story unfolds, it quotes some of the information contained in the messages given to us by the Great Master Spirit Teachers. During each session Sara could plainly see and describe the different Great Beings who spoke through her and she also described the things that took place.

Millions of Immortal Evolved Beings have come to the earth for a definite purpose. They have come in different groups in their space ships (UFO'S) from higher realms of life. Some of them have come from planets in the earth's solar system. Others have come from the high etherean firmament worlds. Also, other different groups of Mighty Beings have come from their far-off higher galaxies of worlds and from higher plateaus of solar systems and their worlds.

These messages were received during a period of several years, beginning January 9, 1957, which was my first private session with Sara. To my amazement at that time, a Mighty Spirit Being named "Sethantes" spoke to me, explaining that he was my assigned Master Spirit Teacher and that he would be with me during every future session that I attended with Sara. He instructed me to always bring my (heavy Revere) electronic reel-to-reel tape recorder each time a session was conducted by PO and ADONTUS who were Sara's assigned Master Spirit Teachers. He further instructed me, that each tape session was to be carefully transcribed by me. Then all typed pages were to be safely filed in steel containers for a future time when the transcripts would be placed into books.

The countless billions of Spirit Beings dwelling in their

various dimensions of spirit life must have names for identification, just like earth people must have names for their identification. However, our Master Teachers explained to us, we must not become overwhelmed or awed by their unusual names. That, really, it is only the quality of the subject matter of their message that is of educational value to earth mortals.

The electronic history records in the heavenly libraries contain records of the earth's Hierarchical Government Organizations that have ruled the earth and its heavens. Those records reveal that the Earth's Chief Governing Celestial Hierarchy is composed of myriads of Exalted, Highly Evolved, Mighty Powerful Beings. They are called Nirvanians, Orians, Gods and Goddesses, who have had many thousands of years of experiences in the high etherean worlds and in other more distant worlds. They too, during many past eons of time, once were mortal Beings living upon an earth-type world.

An Exalted Nirvanian, or mighty Chief God, or a Chief Goddess, is commissioned to rule the earth and its heavens during one cycle of time consisting of about 3,000 years. Then another Chief Ruling God takes his place during the following cycle of time. This same procedure applies to all other worlds. Each world has its own Exalted Ruler, called a Chief God. Earth people should distinguish that a "GOD" and the "ALMIGHTY CREATOR" are NOT the same. The ALMIGHTY CREATOR is the CREATOR of all Gods, Goddesses, Lords and all Human Beings, who are HIS Children.

THERE ARE NUMBERLESS COSMIC GALAXIES

The Great Ones explained there are numberless Cosmic Schemes of Creations, known as galaxies, other universes and other solar systems. All of them exist within ONE GREAT OCEAN OF ELECTRICITY that composes our ENORMOUSLY VAST COSMOS. They operate on different graduated electrical levels, or plateaus, or dimensions, and they function harmoniously through universal natural laws of mighty electrical powers.

Some universes are composed of corporeal earth-type worlds. Each has its own separate heavenly counterpart spirit-world, joined onto and located directly above its corporeal surface. They operate together as one unit. Other universes are composed of countless separate groups of spirit worlds, located in the higher dimensions of the vast cosmos. Each group of spirit worlds have their own various shapes, sizes and colors. They are more highly evolved than the corporeal worlds.

The teachers further explained that ALL LIFE is electrical vibrating energy-power, manifesting through varying degrees of different frequency rates of vibrations. Each human person is an individual universe of atoms, generating his own electrical force-field of energy-power, called his "aura" which surrounds him. Electrical energy power is the LIFE FORCE within all things. Each human person, they said, is a POWER BEING who is endowed with all the universal creative energy-powers which exist within himself. And all those powers must gradually be developed during all of his long journey through eternal life. Thus they explained that ELECTRICITY IS SPIRIT.

EARTHLY VIBRATIONS ARE SLOW

The Mighty Ones also say that the earth and its inhabitants function through slow, coarse vibrations. All Highly Evolved Immortals function through faster, finer, more subtle frequency rates of electrical vibrations, which exist in their refined vast realms of spirit life in higher worlds of different texture from the earth planes.

When the Immortal Spirit beings come to the earth and usually are unseen by mortals with their physical eyes, it is because mortals function through their slow earthly vibrations. Spirit Beings and mortals operate on different rates of vibrations. However, there are many mortals, who, at times, can psychically see (with their spiritual sense of sight) Spirit Beings who are on the earth. They can also see phenomenal things which now are becoming more prominent. A great number of such spiritual things which at first may seem strange and unbelievable, eventually will become commonplace during this NEW AGE.

As our solar system of worlds travels through the Etherean Spiritual Arc of Kosmon, the GREAT SPONTANEOUS SPIRITUAL KOSMON LIGHT is radiating and flooding its rays over all those worlds. That LIGHT is spiritually awakening the MINDS and the spiritual senses of millions of people on the earth, during this present NEW AGE CYCLE OF TIME.

Before this GREAT SPIRITUAL LIGHT can harmoniously operate and fully manifest spirituality on the earth, the earth must first be cleansed of all its evils and iniquities. Therefore, the earth now is going through its cleansing period of time. Then, after its cleaning is complete, all earth people also will have become changed from unrighteousness to righteous ways of living. At that time, all people then living on the earth and in all of its heavenly planes will freely function through spirituality and not mortality. Thus, they will SPIRITUALLY LOVE ONE ANOTHER.

PART 2

EXPLAINING THE ART OF MATERIALIZATION

MATERIALIZATION SESSIONS QUOTED

THE MASTER JESUS, OUR ELDER BROTHER

TES, THE GREAT MASTER SPIRIT HEALER

EXPLAINING THE ART OF MATERIALIZATION

Many predictions were given to us by the various groups of Great Spirit Beings who came to our little center from their far-off worlds. Some predictions are lengthy explanations and others are brief statements.

My Spirit Teacher, Sethantes, who was the First Chief God assigned to rule the earth, 73,000 years ago, said to me: "There is a great work ahead of all the Mighty Ones who have come to the earth. All of them are preparing a COLOSSAL WORK such as your mind can not fathom at this time. These New Age Teachings that are now given through this channel (Sara) are of FABULOUS KNOWLEDGE that has never been brought to the people on the earth. In fact, the very things you are being told about definitely will take place. You will find there have been no mistakes made in what is given here at this center. All predictions and other information given here will come to pass during some future time." (That message dated 9-28-60).

The many messages brought to us are very far-reaching. Their intricacies are almost overwhelming to me. It is impossible to try to cover all of their various topics in this book. Therefore, I will mainly explain what happened during special materializations that took place in the privacy of Sara's modest home. I will also quote astonishing predictions.

* * * * *

When we first started receiving these mysterious messages, we were puzzled and perplexed as to just WHAT was taking place. We could not figure out WHY we were selected for such a profound work.

To help us verify the authenticity of the messages given to us, Sara arranged for a private materialization session to be conducted in her home. We wanted to find out to our entire satisfaction if some of those Spirit Beings, who were speaking through Sara, actually were the same Personalities who perhaps might be able to fully materialize themselves before us.

Therefore, during October, 1957, Sara invited a local medium, named EDITH NILES, who was a very powerful materializing medium, to come to her home to conduct our first private materialization session. Edith's Alchemist Spirit Doctor, named DR. WEST, was well skilled in conducting materializations.

OUR MASTER TEACHERS APPEAR

We were happily surprised and elated when several of our Master Spirit Teachers took their turns and appeared before us fully materialized. They were vibrant living Personalities, appearing in their spirit bodies, displaying themselves in normal physical-type human body forms and attributes. They freely talked with us in their own personal direct voices. As they touched us on our arms, while walking around the room with us, we could feel the warmth of their spirit bodies. We also very plainly could see them eyeball-to-eyeball. Each one of us in our circle-group clearly saw all of the same things that took place, seeing the same details each time during the first materialization and all other following materializations.

Sara and I always had an afternoon session preceding the evening materialization. Often a Master Spirit Teacher would say: "I will try to manifest myself tonight if the conditions are favorable." Then, to our delight, when he did appear and say: "We had such a nice meeting this afternoon" that proved to us beyond any doubt, that we were receiving authentic messages from the Spirit Beings who had spoken to us that same afternoon. My Spirit Teacher

"SETHANTES" made such a *remark* on January 2, 1959, and also on May 8, 1959. On April 29, 1960, the Great Spirit Being named "ROCK OF AGES" materialized and said to Sara: "Thank you Sara, for letting me use your highway of flesh to come and speak to the mortals who attend your private sessions."

Oh, what great joy we experienced when the Great Ones fully materialized themselves before us, each one expressing his own individual vibrant personality! They were clothed in their chosen, especially created, elegant magnificent robes and other styles of garments, which were made of silver or gold or other different unusual types of materials. Their garments were adorned with brilliantly shining precious gems, such as diamonds, rubies and emeralds. Most of the Spirit Beings wore odd-shaped headdresses, which were decorated with jewels, signs and different symbols, denoting to the other Spirit Beings the far-off spirit worlds from which they came. Those symbols also denoted their rank and station as Mighty Gods and Great Master Teachers. Several Spirits wore jeweled shields, or jeweled lapels on their garments.

Their jewels represented different energies and rays and not our earthly concept of jewels as wealth. In the dimensions of spirit life the Spirit Beings do not need things of wealth. However, they quickly can materialize jewels and many other things from "out of the air" from the etheric atoms existing in the ethers of space. They create their garments and other things merely by using their POWERFUL MIND and CREATIVE THOUGHTS. Their personal direct contacts with us and the quality of their messages were the important part and not their jewels.

However, we soon understood that these magnificent gems *symbolized* the beauty, color and light-energies abundant in the etheric realms from whence they came.

All materializations were conducted in Sara's home, once each month. They began Oct. 25, 1957 and ended May 27, 1960. They started at 7:30 in the evening and lasted for about one hour, while Edith Niles was in a deep sleep trance.

Each time a materialization was scheduled, I prepared

Sara's living room for the night seance. I covered the three windows and the front door to keep out all street lights. Then I prepared a so-called corner cabinet. I strung four separate long black cotton curtains onto a long pole and placed the pole across the top of the cornice boxes of the two adjoining windows, with the curtains hanging down to the floor. A plain straight wooden chair was placed behind the curtains, to be used by Edith while she would be in a deep trance during a materialization session. On top of the hall door frame I hung a red-ruby photographer's light that had an adjustable metal shield. That was the only light used in the room and it shone across the room upon the black curtains. Our eyes always quickly became adjusted to that light and we could very plainly see the fully materialized Spirit Pesonalities when they came out walking from behind the curtains and in plain view, stood before us in the center of the room.

A very sensitive microphone which was attached to my tape recorder, was hung at the front top of the cabinet, at the center opening of the curtains. It recorded the direct voices of the Spirit Beings who appeared before us. It also recorded all other voices in the room. I have all of the materialization tapes and their transcripts.

During every seance all outside doors to Sara's home were securely locked and all bright lights were turned off.

Edith's Spirit Dr. West controlled every materialization. Also, Edith's loyal American Indian, named Red Blanket, would firmly hold her physical body while she was in a trance during a seance. He explained to us that he gets inside of her body when he holds her.

For the benefit of readers who are not familiar with the art of materializations, the following explanation may be helpful.

Edith's Spirit Dr. West explained to us that every earth person exudes an invisible psychic gas from all their body openings and from their solar plexus. The Spirit Teachers call it "sensagall gas" which is contained in a little sac, a ductless gland, located at the side of the solar plexus and at the end of the cardiac nerve. The mortal doctors on the earth call it the "duodenum sac." The gas is electrified magnetism, an energy substance used in the production of all kinds of psychic phenomena. Earth people generally call it "ectoplasm."

A materializing medium is gifted with a great amount of ectoplasm, enabling her or him to produce the psychic phenomena called materialization. Spirit Personalities who fully materialize themselves, usually use the ectoplasm to cover their semi-solid spirit bodies, to enable them to make themselves visible to the physical eyes of earth people. The Spirit Co-Workers who assist Dr. West, gather the ectoplasm that is exuded by the people in the room. Then they add it to the ectoplasm produced by the medium, placing it behind the cabinet curtains. That is the purpose of the cabinet, within which the medium sits. The cabinet is used to garner the ectoplasm.

As the ectoplasm exudes from the medium's body, it forms into a shaft-like form, which remains attached to her, adding it to the ectoplasm in the cabinet as it is formed into a big blob. Harm can be caused to her if a mortal person embraces a materialized Spirit Being. Such an embrace by a mortal's magnetic flesh body, reverses the electrical charge contained in the medium's ectoplasm, thereby sending a sudden electric shook back to the medium's body. However, such a shock does not seem to happen when a fully materialized Spirit Being touches or embraces a mortal person.

When a materialized Spirit does not return to the cabinet and quickly disappears, he merely discards the ectoplasm covering his spirit body. He then becomes invisible to the physical eyes of mortal people. Then the ectoplasm is quickly gathered by the Spirit Helpers and returned to the cabinet, to again be used by other Spirits for their manifestations. When there is sufficient ectoplasm in the cabinet two or more Spirits can appear at the same time.

MATERIALIZATION SESSIONS QUOTED

To explain more clearly the outstanding events that took place during the materialization sessions when the GREAT SPIRIT BEINGS appeared before us, all in the group were well acquainted with their own particular Teachers but had not seen them in names to identify them, they are referred to as MASTER TEACHERS.

They walked out from behind the cabinet curtains and stood in the center of the room. Then they lovingly greeted us with their direct voices. Our Master Teachers would call us by our names to come forth and stand before them. Dr. West always conducted the materialization sessions.

Our first session was on October 25, 1957. Dr. West greeted us, saying: "Can you see me? I have to come out to see about the vibrations in the room for my medium (Edith) who is in back of the curtains. Now, if you will pull aside the curtains, you will see her there in a trance."

PO, one of Sara's Master Teachers, appeared and stood beside Sara. He said: "This is the Golden Hour." Then he walked around the room, saying: "I want the other people here to closely see me." Then he returned to the cabinet. (PO was born in China about 6,000 years ago.)

OSIRE (The True OSIRIS) ruled the earth as its Chief God, 12,100 years ago. He said to Ray: "Can you see my features?" Ray: "Yes, I can." OSIRIS: "This home shall become a beacon light watchtower and it shall help others. I say to the group, I have been here many times before tonight and I want you to know and to realize how close we are together. Come, Ray, I will go back to your chair with you. Time is of the essence, as you are regulated by time clocks, but we are not and we give no thought to 'time.' " Then he dematerialized himself before us.

THOR, who was the Chief God who ruled the earth 15,400 years ago, said: "I want you to see me face-to-face. You all may come up closely to see me. Sara, I will take your arm and be glad to escort you across the room to your chair, come, let us go together. I must go now and you all may be seated."

Several other Spirit Beings appeared during this very short session.

On November 29, 1957, ADONTUS appeared and said to Sara: "My child, I want you to know that as the gentle breezes of time pass over your life, your path will be unswerving. I also want to greet the group. I am not using much power because of the vibrations of the outside wind storm. Can you see me?" Sara: "Yes, I see you." Then Adontus said to the group: "You are all children of the CREATOR'S Kingdom. HE speaks through the teachers here at this center." Then he disappeared from our sight. (Adontus and PO conduct all private sessions in Sara's home. During a private session on June 17, 1957, Adontus said: "I have been in the spirit worlds for more than 50,000 years.")

SUE, one of Catherine's Master Teachers, said to her: "Everything is eternally right in the CREATOR'S Hands." Then he said to Sara: "Great spiritual truths will come through you, dear Sara." (Sue was the Chief God of the earth 21,000 years ago.)

RED BLANKET said: "Wind blow so hard that Dr. West he say he can't glue the Spirits together fast enough." (All outside doors and windows were tightly closed, but the wind storm affected the materialization).

SETHANTES, one of my Master Teachers said: "Come forth, Lara, my child can you see my face?" I said: "Yes, I can plainly see your face, and you are beautiful." He replied: "I want to be beautiful in spirit. The CREATOR blesses you. Carry on and do not ask for a lighter task, but ask for a stronger back to carry it. Keep your thoughts high and you will overcome all difficulties. You have been given the TRUTH at this center and no person can ever take it away from you. Now I am going to disappear from in front of you."

(Sethantes was the First Chief God who ruled the earth 73,000 years ago).

DR. WEST: "The medium must come out of her trance immediately, because the wind outside is blowing with great velocity."

Later, the Teachers told us that when they come to the earth from far-off worlds, they are affected by the earth's disturbing weather conditions.

December 27, 1957, Dr. West appeared and examined the vibrations in the room and checked the red light to make sure it was properly adjusted. He said: "The Teachers come to this center to work."

ADONTUS said to Sara: "Do you see my eyes as you look into my face?" Sara: "Yes, I see your eyes and I see the jewel in your forehead." Adontus: "Oh, I thought I had omitted it. Each one of you are here on earth for a purpose and when you fulfill that purpose you shall journey on. Always have patience and tolerance as that is how a person grows in spirituality."

DR. WEST: "Adontus looked just like a sunburst. He was exquisite."

SUE, special Guide to Catherine, said: "My child, I place over you my mantle of protection as you stand here before me. I want you to stand straight, because when you bend

---

DR. WEST—Spirit-Guide from an Invisible Realm.

Dr. West played a particular role—that of door-keeper/control/—guardian for Edith Niles, the medium, and for Sara and her group when in session together. Hence all low or intrusive elements could be kept out.

It was one of Dr. West's special functions in each materialization session, to describe in detail what the Beings looked like and what they wore, as the experience all were sharing *visually,* needed to be recorded audibly in WORDS to be picked up on the tape recorder. These highly important tapes, as stated in the Introduction were the corroborating records of the astounding events taking place, and were to be stored for future transcription into book form. That is how the reader of the messages that follow, is also able to visualize these sessions, due to Dr. West's and Sara's descriptions on the tapes.

forth, I cannot draw power from your solar plexus. I am a little taller than you." Then he walked over to Sara, saying: "You are the sage of this center. Always expect the best, for whatever a person sends out into the ocean of life, it always comes back into his own sea of life. The law of attraction will attract many wonderful things to you."

SETHANTES said to me (Lara): "Do you see my hair?" I replied: "Yes, it is black." He said: "Remember, I am always with you. Every experience that you have you need for your own soul's unfoldment. Now, I will escort you back to your chair." Then he took hold of my right arm and walked with me to my chair.

PO said to Sara: "Do you see my jewels?" Sara: "Yes, they are beautiful." PO then explained: "We create the jewels and put them on, then forget we have them. While we are in the cabinet getting ready to come out, the jewels are all white. When we come out and walk into the energy, into the aurical vibrations of you mortals, then the jewels take on their colors. I will demonstrate this to you. As I approach you, Sara, the jewels take on a golden colored hue. As I go over to Ray, they take on a blue-purple color. They get the colors from your aura surrounding you. The colors of the jewels come from the chemistry of your bodies. During these sessions we also draw from the colors in the room, such as your clothing."

We enjoyed a reunion with our departed family members on January 31, 1958.

TONY'S mother and father both appeared at the same time. His mother said: "This is wonderful to be with you tonight. I touched you as I want you to know that I am so very close to you." His father said: "Hello Tony, this is your dad and I am so glad to see you, my son. Hello to everybody here." Tony: "Hello mother and dad, I am so happy both of you could come tonight." Then they disappeared into the cabinet.

Then my two brothers Hank and Herman appeared with their arms linked together, saying: "Here we are, right here together." I said: "Hello to both of you." Then both brothers walked over to where I was seated and they stood so very close to me that I could very plainly see them. Hank said: "Herman, you stand on one side of Lara and I'll stand on the other side. Lara, what is that in your hair?" I replied: "They are two bobby-pins." Then, playfully, Hank took out one pin from my hair and Herman took out the other one. I held out my hand and they put them into my hand. Hank said: "This isn't half of what I am going to do to you." Herman said: "Here it is, that small perfume bottle that I took out of your purse, and he handed the bottle to me." I said: "I am so very proud of both of you boys and I love you very much." Then my mother and father appeared and joined Hank and Herman. Mother said: "Oh, those boys, they are playing tricks on you. They took the perfume bottle out of your purse and put some perfume into the room." They said: "Hello, Tony." Then they disappeared.

CATHERINE'S departed spirit husband "Mack" appeared, saying: "This is Mack. Know that I love you and I am glad I could come tonight." She said: "I do love you Mack."

FRANK, Sara's brother, appeared. He said: "This is Frank. It is a pleasure to come to you." Sara: "Hello, Frank, you touched me on my knee." He disappeared.

EARL, who is Ray's brother, said to Sara and Ray: "This is Earl. Sara, do you know that your cousin Arthur is over here in our side of life?" Sara: "Yes, I know. How is aunt Ida?" Earl: "She is here but has gone to the kitchen looking at that cake." Then aunt Ida appeared, saying: "I just want to say hello. You have a lovely meal in the kitchen that you all will enjoy after this session is completed." (Sara's aunt Ida is the mother of Arthur and Dwight D. Eisenhower).

DR. WEST: "Sara, there is a lady here with a very old thimble and she wants to give it to you as it belonged to her." Sara: "That must be my grandmother Eisenhower." Then the grandmother appeared, saying: "Hello Sara, you know I had a large thimble as I always had large joints in my fingers." Sara: "I am so glad you could come tonight. Come over here to me, as I have my hand out and want that thimble." Then the grandmother gave it to Sara, saying: "It is awful black and you will have to clean it. It is sterling silver, try it on to see if it fits you." Sara tried it on and kept it. Grandmother: "I am watching my grandson Dwight."

PINK ROSE, who is a spirit helper to me, appeared with a little spotted puppy dog and it went directly to Tony and sat at his feet. Pink Rose said: "This is Pinky. Hello everybody. This is not a squirrel that I brought, it is a little spirit puppy dog with spots." This was the first time I had seen a materialized animal pet.

Also, on 9-25-58, Sara had invited a guest to attend that session. His spirit wife appeared, bringing with her their small dog that had died. While ED and his wife talked, that spirit dog joyfully wagged its tail and loudly barked, proving that our pets survive their earth life. They find their former owners and live with them in the spirit world.

A Family Loved Ones Materialization was conducted on 3-25-60. During this joyous session much talking took place with our departed family members, who fully materialized themselves and we walked and talked together. Each Spirit Being expressed his (or her) own individual personality, exactly as they did during their earth life. Our mothers, fathers and other loved ones appeared clothed in garments such as they wore during their earth life.

Sara's mother and her brother Frank brought with them Sara's sister Minnie who had recently made her transition into spirit life.

ELIZABETH GAINOR, who was a trumpet medium during her earth life, had visited Sara many times. She always wore a pair of the old-fashioned type of eyeglasses, having small lenses and gold rims. As she stood before us she wore such a pair of eyeglasses. She said to Sara: "The only reason why I am wearing these glasses tonight is for added identification.

My brother Hank during his earth life enjoyed smoking a pipe. He appeared with a spirit-pipe in his mouth and the pipe had a spirit light in it. He explained that he brought that pipe for fun, as we always had teased him about his many pipes.

DR. WEST: "When Hank came back into the cabinet, he went right through Elizabeth Gainor's etheric spirit-body and she quickly put up her face to protect her eye-glasses."

Jean's Uncle George, during his earth life, when Jean was a very little girl sitting on his knee, he played nursery songs and other tunes on his harmonica. He appeared, then quckly materialized a spirit harmonica. Then, very loudly he played the tune "Mary had a little lamb." All of our group clapped our hands in tune with the music and had fun.

That was a most unusual, outstanding demonstration. I have replayed that tape recording several times for my friends, but must tune down the loud music to avoid disturbing my neighbors. It is difficult for people who hear this tape to believe that the music is played on a spirit-harmonica by a Spirit Being.

Our session on 2-28-58, was a windy night. Dr. West said: "Due to the velocity of the wind, it takes a little longer for us to take the medium completely into a deep trance."

RED BLANKET speaks broken English. He is a lovable person and his humorous remarks were enjoyed by all of us. He said: "Me got medie now where we want her. You know, me old Indian and me hold medie and me nail her to her chair and me control her body at all times."

PINK ROSE, my jovial Spirit Helper pretended to me that she was a small girl, usually speaking in a childish voice. However, I believed she was an adult Spirit Person. Tonight, she appeared as a little girl and in a childish voice said: "I am a big girl." I replied: "If you are a big girl, then grow-up right now here before all of us." She said: "Do you want me to grow this way?" She quickly expanded and extended herself upwards to about five feet as she stood before us. Then Pinky laughed and disappeared. That was an unusual demonstration.

ENOCH, one of Tony's Master Teachers, had spoken through Sara several times during Tony's private sessions. Enoch said: "Tony, I want you to know that I have always been with you prior to this time and I shall always be with you." Tony: "I have been looking forward to meeting you and I welcome you."

SETHANTES said to me: "Lara, I want you to know that I am with you and nothing shall thwart your unfoldment. We have encircled you with a protective shield. The day will come when you will be able to minister to the human family. I want you to consolidate the pages of your transcripts and place them in proper form. They are to be used for future lecture work and for books, later to be used by many mortal teachers."

GOD OF THE DALMATIAN WORLDS appeared. He wore a big breast plate decorated with jewels. He said: "I am the God of the Dalmations and I am helping each one of you. Also, the GOD of the PYRAMID WORLDS is here. I met him in the cloistered corner of the cabinet. I bid you adieu."

DR. WEST: "The wind is blowing up again and Red Blanket, you hold onto the medium."

Red Blanket: "Moti, a nice Indian Princess, she comes out soon to Sara. She has to wait, that wind tears down the work soon as we build it up."

MOTI appeared with long black braids, saying: "I had trouble building-up, getting ready to appear."

RED BLANKET: "I do the best I can and Dr. West is working hard up above on top of the cabinet and everybody else is working hard because of the chemical changes. Everybody, you mortals, had heavy dinner tonight, your stomachs are full, so we cannot draw much energy from you. Now an Egyptian Prince is here and he got his head all wrapped up, like him got sore-head."

MASTER ZERMA, my loyal Spirit Door-Keeper, said: "Can you see me? I can hardly hold together, because of the wind. I want you to see my turban." Then, with his hand he flipped up the lower part of the turban so that I could see the length of it and I could plainly see him.

ADONTUS appeared and Sara asks: "That picture of a Spirit Being that appeared on my ordination photo (6-15-50), is that you, Adontus?" He said: "Yes, it is me, as I was there to bless the ceremony." Sara: "That session we had this afternoon with Lara was so outstanding." Adontus: "Yes, it was beautiful. All of you be true to thine own self. Good evening."

RED BLANKET: "Me now release medie and I will go home with her to protect her."

At the beginning of our session on 3-28-58, Red Blanket said: "My medie she yawn, then me jump into her body. I say a Great Spirit is here and him have lots of power. He say he come like a redeemer and he come from ancient times. He build up now while medie's body gets chemicalized."

Sara: "You are absolutely right, Red Blanket, as this is what he told us today during our afternoon session."

GREAT REDEEMER speaks: "I am the Great Redeemer who again is on the earth plane in spirit form. I greet each and every one of you. I appear before you for the first time. I was of Arabic descent and I lived on the earth in the time of Moses. I am the one who redeemed the people whom Moses tried to bring through the wilderness. I am the First Master Jehovah (or Jehovih). There are many Jehovah's and I am the first seed of them.

"I always wear my amulet arm-plate and my breast-plate and I want you to see them with their jewels. Know that each one of you are a part of the great cosmic universe. Everything that is here on the earth is in the oceans and in the sky. You have all those elements within yourself. You have the creative ability. Your mind thinks in pictures, so first you must have the picture in your mind. Like the hieroglyphic picture images which we used when we taught and scrolled upon the Babylonian walls, depicting many things. I am the redeemer of the human mind as well as the human race. I come to the minds of all people. I am a universal worker."

SARA: "We are truly honored by your presence."

ORSIS, who is not the God Osiris, said to Catherine: "I was the bodyguard to the Queen of Sheba as she sailed down the Nile River. I was her consort when she went to visit King Solomon in his temple. My name is but a label. I wear my turban, can you see it? I invite you all to come forth and look at me." He disappeared.

RED BLANKET: "His jewels made my eyes water."

ALEXANDER THE GREAT appeared, speaking very loud, as he waved his arms: "I raise my arms in protest of all that is evil." (He was born in 356-B.C. and was King of Macedon).

GOD OF THE PYRAMID WORLDS made his first

appearance. He said to Sara: "This place, this ground is magnetized and it is magnetic." (Eight other personalities appeared). During most of the times when our materializations were conducted our Master Teachers appeared before us. However, due to lack of space in this book some repetitions will not be quoted.

SETHANTES on 4-25-58 said: "I spoke to you this afternoon through Sara. You know we teach you truths. I have a different robe on tonight."

SUE wore a beautiful gown covered with jewels. He said: "We have created our garments for this special purpose, to come down into this dark, dank, murky atmosphere. In our minds we created these things. I have my golden sandals on and arrayed myself as you can see."

GOD OF THE PYRAMID WORLDS appeared with his head swathed and he wore a beautiful jeweled breast plate. He said: "I bless this channel, Sara, through whom I speak to you. I look upon each of you as a beautiful soul."

RED BLANKET: "He had his head all wrapped up."

FRAGAPATTI, who ruled the earth as its Chief God, about 8,900 years ago, appeared. He said: "I clothed myself in metallic cloth for protection aginst the earth's vibrations." Then he spoke to Sara: "I want to touch you with my garment and you can almost feel its vibrations and I want you to feel it. Place your hands with palms down in front of you and I will bless you with the electric cloth of the garment."

DR. WEST: "He was magnificent. Did you notice the wireness of the texture of the cloth? I have heard him lecture over here in our spirit world and he has a very brilliant mind."

SARA: "I felt the gold cloth. He is one of the Seven Golden Age Masters who now are working on the earth."

THE FIRST JEHOVAH: "YAHWEH that is the way they pronounce my name in some dialects. Do you see my long beard?" (It was very long—almost to his knees). He said: "Bless you peoples of the earth. The New Age is here."

MOSES appeared. He said: "My commandments were given to me by Great Teachers from the world of spirit. The greatest of the commandments is: "That ye love one another as the CREATOR loves all of us." Then Moses speaks in a foreign language.

DR. WEST: "Moses tells me that he addressed all of you in Arabic, but I did not understand it myself."

SETHANTES appeared on 5-30-58, wearing a beautiful gold band around his forehead. He said: "Greetings to all of you and to my beloved channel Sara through whom I can speak. Remember, that the greatest events in history make no more noise than the rising of the Morning Star. The greatest events are the silent ones. Lara, my special one, may I assist you back to your chair?"

STAR BEING from the Pleiades solar system, said to Ray: "Can you see my headgear that I made especially for you tonight?" Ray: "Yes, I see it and it is beautiful."

MATHENO, who is one of the Seven Golden Age Masters, appeared. "Good evening my friends and to you Sara, my child. I come from the Golden Age and I am helping to bring the Golden Age to this earth world. I come with my gold gown upon me that you may realize that we are now living in the Golden Age. This center is a nuclei of power. This center will go down in history as the starting point of the Great Masters appearing upon the earth, coming down to the earth from the past ages of Time. I lived many centuries ago upon the earth and walked and ate as you now walk and eat. I have returned to walk with earth people today. You, Sara, can see me on occasions." Sara: "Are we doing the things we should be doing?" Matheno: "If you are doing things you should not be doing, then we 'block' everything. There is a continual flow of dynamic power in this house and I see this house as a 'net.' I bring to you the message that you are all ambassadors of the Golden Age. Talk and live the Golden Age, then you live correctly. May the powers of the ALMIGHTY CREATOR rest upon you.

"I come from the Pyramid Age. I sat in the catacombs and I watched the River Nile. I was the captain of one of the great ships there." Sara: "Was it a space ship?" Matheno: "Yes, a space ship which will one day return to the earth."

DR. WEST: "He wore a gold robe and he had a stole made of camel-hair that was exquisitely wrought in texture and design. Another Golden Age Master named KASPAR, is here tonight, performing service in the cabinet, He is electronically helping us."

GOD JUPITER, from the Planet Jupiter (One of Catherine's Teachers) appeared, saying: "This is Jupiter. I have my neck-robe on as I always wore it in the century that I lived in while on the earth. I want to bless you Sara, our channel through which our words flow as we express ourselves. Dear Catherine, I always am close to you when you are giving your lectures. We always stand and encircle you with the radiation of power and sometimes we have to throw a protecting magna-vox about you to protect you from the thoughts which we do not like."

DR. WEST: "The WATCHMAN next will appear. He wears what looks like a badge and he has a staff in his hand. It looks like a shepherd's crook. He says that he protects this house."

WATCHMAN appeared: "I am anxious to reach the mortals that I protect. I use this crook-staff to protect this place as it is my electric wand. I wave it in-and-out of sight."

DR. WEST: "The Masters are busy building up the chemistry in the medium's body."

RED BLANKET: "Master Teacher JESUS comes now. He touched me on my head."

PO, on 6-13-58, appeared wearing a beautiful jeweled robe with a jeweled collar around his neck which extended down to the bottom of the garment. He said: "I am happy to be here in front of all of you." Sara: "Thank you PO for coming. Your garment is very beautiful. I enjoy those talks with you over the telephone." (At times, when Edith and Sara talked on the phone, the PO would come in on the line and directly talk to Sara).

URILLIUS appeared. He is a powerful Master Teacher who came to Catherine. He said: "I told you that I would come so that you could see me. I want you to have confidence in us and know that we walk by your side. I want to ask this community group to keep it intact. There are no heights so high that you cannot scale them. We want you to grow spiritually stronger and unleash that POWER that you have within you."

GOD OF THE PYRAMID WORLDS appeared, saying: "I at one time held teaching circles in the catacombs and also in the Pyramids, which we entered by a secret door. I bless you one and all."

DR. WEST: "He wore solid silver shields as lapels for the white coat he had on. They shimmered with light."

MEGESTE, a Golden Age Master appeared and he was magnificently beautiful. He invited all of us to walk up to him so that we could plainly see him. He said: "I had promised you that I would come tonight and I am here to greet you. I am at this time working over the entire country of China and especially over Manchuria." Then he spoke in the Chinese language, bowing gracefully as he spoke and then he disappeared.

DR. WEST: "I did not understand his Cantonese words, but the interpreter with us quickly vibrated to me that he spoke Chinese. He blessed everyone in this room which he called the 'Temple.' His headdress was very tall and it had jewels, with a flat crown on top of it."

SUE appeared 8-22-58, saying: "I want to thank this wonderful channel, Sara, who is the bridge we cross to build the power at this center to bring our messages."

STAR OF THE EAST appeared with a beautiful brilliant star in the center of his forehead. He said: "There is a great dawn coming over the earth. It is going to touch the hearts of many people."

GOD OF THE PYRAMID WORLDS: "I studied and lived in the catacombs which are the subterranean places of the Pyramids. You mortals can reach the apex of all things." I said to him: "That was a wonderful message you gave to us this afternoon, telling about the Pyramids."

KASPAR, a Golden Age Master, who was a Persian appeared, saying: "I have come to strengthen the power here and each of you can demonstrate this Great Universal Power. Remember, the camel is a wise animal. Like him, as long as you live on the earth you will have to cross the burning sands of experiences."

SARA: "It looked like he had beautiful jewels shining

about his face."

DR. WEST: "He wore a sort of Grecian outfit with bands of Roman lettering that decorated the costume. Its texture was like a silver mist speckled with gold. The jewels you saw were not on his face but they looked like it. He wore some kind of decoration upon his head that was decorated with jewels. It looked something like a veil as it fell alongside of his face."

APOLLO, a Grecian Golden Age Master, ruled the earth as its Chief God 18,200 years ago. He had spoken through Sara several times as one of Ray's Teachers. As Ray stood before him, he said to Ray: "What stands facing you is beautiful, because it is the work of "Spirit" . . . which is the work of the "CREATOR." Then he walked with Ray to his chair and disappeared.

DR. WEST: "His robe was a silver-blue and he wore a pure white scarf around his shoulders. The beautiful cascading jewels were on the side of his robe which was gauze-like in texture."

MATHENO, a Golden Age Master appeared saying: "I come from the world of spirit. Each one of you and all other persons are precious jewels in the CREATOR's Diadem. May the Spirit of the CREATOR always enfold you."

DR. WEST: "He had a closely cropped beard. His costume was silver-white with glistening patterned colors of pastels, blues and yellows, His headpiece was some kind of a veil, like Kaspar, the Persian, wore."

The Golden Age Masters who appeared brought the VENETIAN from Venus, who is a very powerful Being. He appeared, saying: "I speak with the tongue of love. I come because my Planet Venus is driving the earth into a retrograde position. You are now on the horizon of the dawn of a new and better day. When you have love in your heart you are filled with the Holy Power. Life is a great stage and all humans are the actors in the drama of life. You earth people have encountered the opposite of love and you have gained a lesson. You will always encounter the opposite of everything. There is the equal polarity between the peoples as there is between the planets. The planets too, move on into their orbits. On the earth you are held there by gravitation.

When that gravitation is set into reverse, then you are levitating through the powers of resurrection, into the world of spirit. The Planet Mercury rules your intellect, but the best of all is to be ruled by love."

DR. WEST: "As I looked at him, his gown was blue, speckled with pinkish-red spots. It was beautiful in its texture."

SETHANTES on 9-26-58, said: "Do you see my station insignia?" I replied: "Yes, and it is attractive." I asked him, "Did you strongly touch my head yesterday while I was typing?" He said: "Yes, I wanted to lead you into another vibration where you could do your work more efficiently. Also, at night when your physical body lies in slumber, I take your own Spirit, in its etheric spirit-body and you join me in the astral regions. You will never permanently come over here to the spirit world until you are disconnected from your physical body. You travel out at night from your physical body by the extension of the psychic umbilical cord. The more you are growing spiritually the more elasticity that cord has and the farther we can go into the temples of the astral."

GOD OF THE PYRAMID WORLDS stood in front of the curtains. Then he sang loud in a melodious baritone voice, these words: "He walks with me and he talks with me and he tells me I am his own. And the joys we share as we tarry there, none other has ever known." We applauded his singing as his song was one of our favorite songs that we sang during our sessions with Sara.

Then he said: "There are many Gods. The Pyramids represent strength and depict the growing step-by-step of the creative process of the apex. Every person can be a Pyramid unto himself. That was I singing. I used to sing in the temples. I will sing again sometime. Thus, I build up my spiritual matrix voice."

DR. WEST: "His song 'In The Garden' was beautiful and his robe was elegant."

PYTHAGORAS was a Greek Philosopher, born in Samos, Greece, in 582-B.C. He said: "Come forth, the hosts of this house." Then Sara and Ray both walked up to greet him. He said: "I lived over 500 years before the birth of our Elder Brother Jesus." Then we walked around so that all of us could plainly see him. Saying: "I am glad to have the privilege to adorn this temple-house with the rest of the Masters." Sara said: "Thank you for coming and do come again as we welcome you."

DR. WEST: "PYTHAGORAS is a Great Master and he wore a different color and contour of a robe, with a court collar of his kingdom around his throat."

LEMURIA appeared on 9-26-58. He said: "My name is LEMURIA, which is but a label as I speak the Arabic language, which you do not understand. I greet you Marian and the rest of the class. Each of you have a candle so let its light shine forth from the hilltop of your expression."

(During sessions with Sara as the channel, he brought other messages to us telling us about the submerged continent of Lemuria).

QUEEN NEFERTITI appeared. She was the Queen wife of the Egyptian Pharaoh named Akhenaten. She wore the same style of headdress like she wore while in earth life, often shown in paintings of her. She also wore a wide collar that adorned the lovely neckline of her dress. She was very beautiful. She said to Catherine: "I help you with your lectures. I want to thank you, Sara, for permitting me to speak through you during class sessions."

SUE appeared on 11-7-58, saying: "Honesty is the best policy and what you send out into the lives of others will come back into your own life."

ENOCH appeared. He said: "My robe depicts my station of life. It has lapels of solid silver circles. I invite all of you to come up to me and look at my silver lapels. These have been apported from the earth world to me over here in the spirit world. They have initials on them."

RAMESES, the Pharaoh King of Egypt, about 1315-B.C., appeared. He said: "Part of my gown is made of camel's hair and I want all of you people to see it. May the Power of the GREAT SPIRIT forever be with you."

DR. WEST: "His entire gown was not made of camel's hair, but what he had around his neck was a stole that hung down the front of it, that was made of camel's hair and the rest of the gown was white."

On 12-5-58, SETHANTES walked with me around the room, saying: "Our session this afternoon was heaven-sent and I thank you, Sara, for permitting me to speak through you." I said: "I feel you touching me on the top of my head many times." He replied: "Yes, I do touch your head because hair contains more magnetism than any other part of an earth flesh body."

VINYU, a dancing girl came to Jean, appearing in a beautiful dancing dress having many folds. She was very graceful as she danced for us, with the many folds of her gown that fanned out in billowy waves of softness and her jewels brightly sparkled with their many colors. She said: "We can't stay too long in the vibration because it takes too much power." We applauded her performance. She said: "I danced in Solomon's Temple when I lived upon the earth."

SETHANTES on 1-2-59 explained: "We are chemicalizing and magnetizing the bodies of all this group. We take our magnetism to convert it into ectoplasm. We use this in our manifestations here before you."

ORLANDO, a Great Master from Venus, appeared, saying: "I come from the vibrations of Venus. I bring to you the message of love and I bestow upon all of you the vibrations and radiations from my planet. You can live on Venus even though you live in a physical body, because you are nothing more or less than 'MIND.' When you inject into your mentality the divine vibration of Universal Love, then your life will be on earth what it was originally intended to be. A life of usefullness, of selflessness and a life of love. DIVINE LOVE IS THE 'KEY' THAT WILL UNLOCK ALL THE MYSTERIES OF LIFE."

DR. WEST: "You could see the radiation from his power. The band he wore around his neck was like a silver ruff. He wore his jewels down the front of his garment."

SARA: "He has spoken through me many times."

ENOCH, on 2-6-59, said: "Can you see my new robe? I made it this afternoon. Sara, dear one, I was weaving this robe while I was using your mentality of MIND. I want all of you to see it and also see my surplice (outer garment) that I wear over my robe when I come from the ninth plane. As you know, there are other higher planes of spirit life which surround the earth." His robe was intricately and finely woven.

LEMURIA appeared: "I came from the far-off Spherical World. When I lived on the earth I lived on the continent of Lemuria, about 33,000 years ago, before it went down in the Pacific Ocean. My message is: 'That the "All-Seeing-Eye' which is within you, is always upon yourself."

FRAGAPATTI appeared, saying: "I want you to see me and my new jewels that I have created. My jeweled collar that I am very proud of contains all the colors of the different rates of vibratory energy."

SETHANTES appeared: "There are myriads of worlds. And in the great domes of the solar systems there are no coverings. There is no space that can be defined because it is endless. There are planets out there in the universe that are drawing close to your earth. They are peopled with intelligent Beings. The space ships and those who occupy them are already close to the atmosphere of your earth and they are now penetrating the earth. That is what it will take to "knock the bottom out of theology" and clean away all of the erroneous pagan teachings of the past ages. Then humans will realize there are higher worlds than the earth world."

SISTER PLEIADES made her first contact at our center. "My name is Sister Pleiades and I represent the Pleiades Solar System of worlds. I am a spiritual sister to the STAR OF THE EAST, who is one of Ray's Teachers. Tonight, Ray, I touch your eyes with the power of spirit." Then she placed her hands on Ray's head. She said to Sara: "I will come and speak through you because you are a clear channel. I will use your vocal-cords and I will speak inside of you. There is only one thing in all LIFE and that is LIFE ITSELF."

On 3-6-59 SPHERICAL materialized himself before us. He had arrived from the Triangle Worlds on 1-7-59, when he spoke through Sara. He said: "Good evening friends, I am Spherical. Can you see my gown and crown? I come within the sphere of your earth life with a radiance of health and power. Many advanced things are coming to your earth in the way of space-lore. There will be space platforms built. They will be held up in space by the law of gravitation. I lived upon the continent Lemuria that was on your earth at one time." He appeared as a large man with a tall crown and he wore a lovely garment.

LIKA, who is a Chief God, ruled the earth during the last 3,400 years. He appeared, saying: "Blessed ones of the earth, I am spreading out my robe so that all can see me and see how I was working with it around you, Marian. It is a robe of security and of well-being. Thank you, my dear channel Sara, for permitting me to speak through you."

DR. WEST: "He wore a plain robe with many irridescent colors."

RAMESES appeared, saying: "I have my desert costume on and my turban. I was one of the Pharaohs of Egypt. My bones are still lying over there in one of the crypts, but I am here. It is wonderful that we can come this way to bless you." Then he spoke in Arabic and disappeared from our sight.

DR. WEST: "He was very powerful and wore a shining headband."

ADONTUS appeared, saying: "Can you see my breast-plate? It is composed of brilliant gems. I use it in my concentration. In our spirit realms we also concentrate the same as you do on the earth. Because all that you have is your MIND and all that we have is our MIND. If we did not concentrate, how could we send a thought to you? Remember, that THOUGHT is the greatest power that you have on earth. It connects you with your Creator. Thought is mightier than the pen and it is greater than the sword. Your voice merely materializes your thoughts. We hear you "Think" and others hear you speak. That is why we vocalize through you, Sara, and produce through your throat-box our thoughts so others can hear with their physical faculties. WHAT COMES FROM US IS CORRECT."

On 4-10-59, DR. WEST said: "Each of you mortals have a Spirit Doctor Alchemist and I am the Alchemist working with this materializing medium named Edith. The Doctors now will come out before you and give you healing treatments."

As the Spirit Doctors took their turns and appeared, they wore earth-type business suits. Each brought with him an electrical spirit machine to use on us during a treatment. When they used the spirit machines we clearly heard the buzzing sounds from those machines.

DR. TAYLOR, who had attended Tony's earth-birth was his Alchemist Spirit Doctor. He said to Tony: "I want you to know that I am treating the nervous system of your physical body. And as I give you a treatment in the lungs, hips and back areas with this spirit machine, let me know if you feel the tingling vibrations. We also are injecting energy electric molecules into the physical body and are taking care of the glands as we go over the body." Tony said: "Thank you Dr. Taylor. I sure felt the tingling."

Then DR. MILLER treated Catherine, saying: "You have been quite nervous. Let me know if you feel the contact and the drawing power of this machine. You will feel it from the top of your head to the bottom of your feet. We also are taking care of the capillary system so that it is more active than it has been." Catherine said she felt the tingling sensations.

A different DR. TAYLOR treated Jean. He said: "We are going to treat your entire body and you so other people will see that she looks like she is turning black. That is the chemical action of the treatment. Now watch her carefully as her dress changes color. First it looks almost white and then it changes to almost black. See how we go around the contour of the body, down the limbs to her ankles and especially up through the sacroiliac." Then he gives her body an adjustment.

DR. GRAY treated Ray, saying: "Turn your back so we can work on it. We are going to put the 'toximeter' on you to check the amount of toxic poisoning in your body. Do you other people in the room hear that machine?" We replied: "Yes, we hear it." RAY said: "I can feel it." Each person in our group clearly saw the lights flashing about on Ray's head. DR. GRAY said: "Now we are going to give you some more of this electricity and we are going to put three hands on you, one on top of your head and two hands on your back. I only have two hands, but the other Doctor has his hand on you, ha, ha."

DR. LOWELL, Sara's Spirit Doctor said to her: "We are going to use this "spinal ray" on your spinal column and it will feel like two columns going down. We will give it to you through your back and legs all the way down. Do you hear the vibration and also feel it?" Sara: "I most certainly do feel it."

I exclaimed: "Look at Sara, she is beautifully illuminated and her arms are glistening."

DR. JAMES WELLMAN is my Spirit Doctor. He said to me: "We will give you a back treatment." Then he took hold of my left arm, at the elbow, giving my arm a quick jerk upwards, and I could feel the snap of the adjustment take place. He said: "Now we will treat the spine. Remember, the spine is the trunk of your body and all the nerves of your body are like branches of a tree trunk, and there are two covers that cover what the inside of the body contains. We are now giving you this spinal adjustment and you will feel it from one end to the other. We also will treat your sacroiliac." Then he took hold of me and gave me an adjustment, raising me up off the floor, saying: "Now your spine is back in place." I felt the electrical treatment on my back and in my hands. (Marian was absent).

SISTER PLEIADES appeared on 5-8-59, saying: "I want you to know that the Pleiades solar system that operates in the great horizon that canopies the sky, brings power to the earth and just so do I bring power to you Sara and Ray. Other Master Teachers are here and we are all working at this center, which is the "core-nucleus" of this Golden Age demonstration."

The TRIUNE GOD made his first contact at our center on 3-13-59, and he has brought many far-advanced messages. He spoke during our session this afternoon, saying: "The great wind-power that exists upon the Triune Worlds from which I come, is like a storm at sea. It blows constantly like a tornado or a cyclone."

TRUINE GOD fully materialized himself and said: "I am the Triune God of the Triune Universe from which I come. I have within my hands and within my power, the triune capacity of all things that belong to our triune worlds. I have walked with you Sara, in the garden here, and I have talked through you. I will bring messages that unfold the 'Mystery of Life' as it really exists throughout all of creation."

SARA: "He had something in front of his robe and it was quite large and round. I have never seen anything like it."

DR. WEST: "That was the 'Life Dial of the Sun.' There also was something unusual about his headdress."

JESUS appeared and said:

"I walk in the hearts of all people."

"The FATHER in you will take care of all things."

"Unfold ye first the Kingdom of the FATHER within you, that all other things may be made whole."

"The FATHER and I are ONE, and all people are with both of us."

"I bless each one of you."

TRIUNE GOD on 6-26-59 said: "I greet each one of you and bring you security, power and vision. I am a Teacher of the private inner-circle class at this center. May the power of the 'HOLY CHRIST' rest with each of you."

FRAGAPATTI said: "I have been with the band of Sara's Spirit Teachers for twelve years. I want all of you people to see my gold robe. It was woven by my thoughts and it is pure spun gold, signifying that it goes with this Golden Age."

SARA: "I see you standing in a gold light every time I psychically see you during a class session."

JUPITER, from the Planet Jupiter, said: "I wear no head-gear, but just my black hair and jewels. I am one of the many Gods assigned to make the necessary momentary changes in your earth's worldwide monetary systems."

ORLANDO, from Venus, on 7-31-59, said: "The thing I wear around my neck is a scarf. It could be likened to a lace material, but it is etheric and is finer than hair and it feels like velvet. Its color is purple."

MOSES appeared, saying some words in Aramaic. He brought a scroll and in English said: "These are my Ten Commandments and I unroll my scroll before you."

GOD OF VENUS, said: "I bring you Divine Love. The Golden Age has been ushered in. Man has a gold mine within himself. Love is the 'Key' to all of your problems and

Love will protect you. My world Venus is coming closer to the earth."

ISAIAH appeared, saying: "I lived on the earth many centuries ago and they called me 'Old Isaiah' the Prophet, and my prophecies have never died. May you all realize your Divine Power." He had a flowing white beard.

RED BLANKET: "His long beard got tangled in medie's hair as he went by."

MELCHIZEDEK appeared, 8-28-59, and said: "Lara, my child, can you plainly see me?" I replied: "Yes, you look regal in your elegant robe. What is that which is shining so brightly on top of your head?" He said: "It is my light. I will not touch you as I am not very solid yet, but I will walk with you around the room. I too, wove my robe and my mantle by the power of thought. Sara, keep right on with your spiritual achievement."

STAR OF THE EAST appeared on 8-28-59, saying: "This house is highly magnetized and before I could appear tonight I had to get through the psychic-mesh of protection that was placed around this house." He wore a star in his forehead.

Many other Great Blessings appeared.

HERCULES bounded out from behind the curtains with great power, with his arms stretched high above his head, saying: "I give to each of you the power of my strength that I bring to you from the world of light."

DR. WEST: "I was suprised as he nearly knocked me over with his power and I had to pull him away from the medium."

Then a GREAT BEING who identified himself as "HERMES" appeared.

(HERMES TRISMIGITUS, the thrice greatest, is the most revered of the ancient Alchemists. The Smaragdine Tablets were the well-spring in which all alchemy had its source. Hermes is identified with THOTH, THE GREAT ATLANTEAN who employed magical, astrological and alchemical doctrines. As THOTH, he helped build the Great Pyramid in Egypt.)

HERMES said: "All you blessed ones, I greet you from the

world of light. Know that you are living in eternity now, so do not try to rush madly through life. Your body is a temple of the LIVING CREATOR and you are now as much of a Spirit Being as you ever will be. You would not be any different if you were to step through the veil tonight, making your transition into spirit life. Your MIND is a motivating power."

I asked him: "I understand that you were "THOTH" in the ancient Atlantean days. Also, that you helped to build the Great Pyramid in Egypt?" He replied: "Yes, I took part and many others took part. I am an Ancient Atlantean and we understood the Atlantean and Lemurian languages. Many of those space people that are coming to the different parts of the earth world now, they at one time lived on Atlantis and also on Lemuria." SARA asked, "Did we live on those lands?" HERMES said: "Oh yes, all of you have lived on those lands.

"The land you now stand on, at one time was an ocean, and the ocean at one time was land. So the earth had to be turned on its axis in the opposite direction. THE GREAT MASTERS ARE WORKING ON THAT NOW. Great and wonderful changes are coming because you are already living in the Golden Age. I want to greet you, Sara, as it is wonderful to be here."

DR. WEST: "It is true that there is an inner earth inside of this earth's crust upon which you live.

"There are so many Spirit Beings here tonight who will not appear before you. Among them is the LAMA BABAJAI."

THE LORD OF ALL LIFE appeared on 10-30-59. "I come to speak to all people on the earth. Know that LIFE belongs to you and that all earth mortals are Gods in embryonic states. You have lived many different lives. There is no 'TIME' over here in the spirit world. Give your services to humanity. Before you can enjoy the thrills of life in the spirit world you must eliminate from your mortal mind all hates, all selfishness, all greed and all jealousy. All of the material things you have in your possession were only loaned to you, even your children were only loaned to you, and your money also is only loaned to you. You can not bring any of your earthly possessions with you when you make your transition into spirit life. Place your money where it will make the most people happy. You must give of the material things and you must also forgive people, as they know not what they do when they hurt other people." He wore an elegant gold shoulder cape.

ALPHA and OMEGA, THE TWINS, appeared on 10-30-59. Alpha came out first, saying: "Lara, keep your eyes on me because my brother Omega is right here beside me." Then Omega said: "Can you see the transformation, as we are transformed over each other? Life is a circle and you are eternally living in an atmosphere of spirit power at all times. Then they stood separately. Omega wore a small bonnet-cap. Alpha said: "I do not wear a cap and you can see that my head is clean-shaven." I plainly saw their unusual demonstration. They brought a great number of messages to me during my class sessions with Sara.

UNKNOWN ONE appeared, saying "My unusual name matters not. I come to bring to you the lessons of life. As you can see, I hold high above my head the TORCH OF ALL LIGHT. My message is: *know that all is well.*"

URILLIUS, on 11-27-59, said: "I thought I would come out first while the power is still warm. I am here in a solidified body that I have been preparing all day for this meeting. I had to change my bodies from the sensitive vibrations to the earth vibrations. It is just wonderful that we can come here the way we do. SUE is now coming out to join me." They appeared at the same time.

UNKNOWN ONE is robed in silver. "May each of you realize that you are not unknown to us. You have lived other lives on other planets. Carry on with your work because the earth is going through an evolution and a process of rejuvenation. Many wonderful changes will happen in various places all over the earth. The Gods of peace are now walking among the earth people, in order that you may in the future have Peace and Prosperity."

FRAGAPATTI was robed in what looked like white ermine. He talked to Jean.

ISAIAH: "I want you to know that the Sages of the past ages helped us while we lived on the earth and just so now I am helping you and you in turn will help others."

MARTINOLA: "I am from the Planet MARS. I just now came through space and I still wear my helmet. This is my first appearance in this type of a body. The Teachers at this center have permitted me to use Sara as a channel so that I can bring my messages."

SETHANTES: "There is a NEW HEAVEN and a NEW EARTH taking place. And the time will come, Lara, when you will make compositions of your transcripts and put them into books. When that time comes we will go along with you. I say to you Sara, I want to thank you for the wonderful meeting we had this afternoon."

On 12-18-59, ENOCH said: "I am wearing a gold and silver robe. It has the basque front like we wore in the old biblical days while I lived on the earth. It took me exactly one minute of your earth time to create it. Tony, I am almost as tall as you and I can even make myself taller." TONY: "Your robe is a most stunning costume." We all saw his stunning robe.

MELCHIZEDEK, my Teacher, appeared. He wore a beautiful jeweled cross, studded with diamonds and emeralds. It was about eighteen inches long.

SETHANTES, my Special Teacher, always was present at every session that I attended in Sara's home. He appeared and said: "Can you see me?" I replied: "Yes, and your robe is beautiful." He wore a wide collarette that was gold, studded with sparkling jewels.

DR. WEST appeared and walked to the place where stood a small heavy white wooden table in the room. He picked it up and carried it and then placed it in front of the curtains of the cabinet. He said: "You noticed that I carried that heavy table with the greatest of ease. Then he again picked up the table and held it up high and carried it back to its place in the room. He said: "It is a levitating force that we use."

On 1-29-60, A Great Master Teacher named "STAR OF THE EAST" appeared. He wore a silver robe. We also saw lights of great power flashing around his head.

Also on this date, SETHANTES appeared, wearing a robe

spun of pure gold shimmering cloth. He told me to touch it and I felt its texture.

DR. WEST explained: "He permitted you, Lara, to touch it because we had insulated the medium for that demonstration." (Dr. West spoke from behind the cabinet curtains).

Then I asked SETHANTES: "When I start writing the messages I have received, is that information important now at this time?" He replied: "It will be of vast importance later on. All of the messages given to you will later be put into book form in the days ahead."

OSIRIS also appeared. He said: "My jeweled robe has a mantle-drape that I can use to cover my head when I cross the Sahara desert." Then he pulled the mantle over his head.

APOLLO, on 1-29-60, said: "I speak unto all the people on the earth. I bless this house and this New Golden Age of Revelations. Remember, that all people on earth are Spirit Beings and that each one of you can do anything that Apollo does. You have within you the Mighty Powers of your MIND, that Great Dynamic Power-Force that controls all the world in the ethers of space. There is not one thing impossible if you just set your MIND to do it." (During a prior session with Sara, he told us that he at one time lived on Lemuria. That his name signifies Love, Harmony, Symmetry and Music).

ZARATHUSTRA was garbed in all white, saying: "Thank you for inviting me into your mortal midst. This center is becoming highly magnetized. I want you to know that the first people who lived on the earth were all materialized Spirits, as there were no mortals. They were created from out of the cosmic rays and they were Souls without any flesh. The mythical biblical story about Adam and Eve is hypothetical all the way through." (He lived on the earth about 9,000 years ago in Persia. The first bible ever given to mankind on the earth, was given to Zarathustra by his Spirit Teachers. All bibles since, have been to some limited extent, restatements of the basic principles and laws embodied in the Zarathustrian bible. He is also known as Zoroaster).

SARA said: "Zarathustra has been with me ever since the year 1932."

NORTH STAR said: "I am the North Star and I greet you

for the first time in this form. I come from the North Star, from the far north. I was known at one time as ORLEANDRO in the ancient days. I helped the white man to discover the North Pole." He wore what looked like white fur.

ISAIAH said: "You are a Spirit and you are a Soul. All people are Souls. An earth person is a Soul pushing a flesh body around like you would push a cart. And as you go into the market places of life, you push your flesh body in which you live. You are just as much of a Spirit now as you will ever be. You won't be any different when you come over here in the spirit world. You will not sprout any wings, which I thought I would do when I entered the spirit world, but I never found any wings. I found that I was the same "Old Isaiah" as I always was. I want to give you wisdom and knowledge as they are the stepping-stones."

On 2-26-60, A Great Being with the name of SPIRIT OF DIVINE LOVE said: "Divine Love is the unseen power and the divinity of man is the expression of that divine love. Who is the CREATOR? HE is the DIVINE OVER-SOUL of ALL LIFE, yet every man unto himself is a Creator, because every person has within the center of his Being, all the creative possibilities to spiritually grow and become a Great Master Teacher."

ZARATHUSTRA said: "I want each one of you to know that you are just simply excursionists on the train of life. You are sitting on the same train together and some day you will come over here and land at the same terminal station of life, so you might as well be friendly as you travel on this journey, called LIFE." He was very tall as he stood before us.

On 4-29-60, ROCK OF AGES said: "Each one of you,—through the evolution of your own Soul, through space and matter,—may become an atlas of your own world. I bring to you the scroll of LIFE, which I hold before you.

"You have within yourself everything that exists in the spirit world of light. YOU ARE A LIVING LABORATORY OF POWER. You have only 'One Power' that you can bring with you when you make your transition into the spirit realms. That power is what you have accumulated while living upon the earth. May the peace of all the planets rest upon your heads.

"GREAT THINGS ARE COMING TO YOUR EARTH BEFORE LONG AND ALL PEOPLE ARE STANDING ON THE ROCK OF AGES.

"Each earth person is an embryonic God unto himself. Beloved Sara, thank you for permitting me to use your highway of flesh to bring messages to the various ones who attend this center."

"I BLESS ALL OF YOU AND MAY THE 'HOLY CHRIST' WHO WILL SOON APPEAR, LIVE FOREVER IN YOUR HEARTS."

DR. WEST: "His robe was radiant and it had all the Colors of the prism, with lights glistening from it." (Rock of Ages made his first contact with us on 4-11-60)

SETHANTES: "When the Teachers appear here before you their bodies are solid."

ORLANDO from Venus: "When a Human Being's Soul is ripened, and his MIND is ready to work spiritually, then the Soul stores its information in its vault."

MELCHIZEDEK on 5-27-60 wore an elegant robe of white satin and silver and a crown of jewels made of red rubies and emeralds. He said: "The CREATOR is within each one of you, in every human being. You must have faith in your CREATOR, Who is your own Higher Self. I bless you, Lara, with my scepter as I touch your head."

SUE said: "I want to show all you people my new robe." Then quickly, he materialized an extra robe and placed it over the robe he was wearing. That was an amazing demonstration.

SETHANTES: "There are space ships all through the universe and through the earth's atmosphere. One is anchored high in the sky above this house. There are many space ships flying around above the earth's continents, but they are invisible to many mortals."

SCARLET PLANET GOD spoke through Sara for the first time this afternoon and he promised us that he would materialize himself tonight. He now appeared, saying: "I come from the Scarlet Planet of this universe and that is the Red Planet which earth people call MARS. It exerts great force and power in your solar system.

"I say to all of you people here, if you want to grow with force and power and become spiritually strong, you must relinquish your hold upon negative thoughts such as jealousy, hate, criticism and greed. When we come to the earth from MARS we feel there are such negative forces that we must contend with, that we come with profound force and power to drive those evil forces out of the hearts of earth people. Good-night to all of you."

DR. WEST: "The materializing medium, Edith, is very powerful. She is doing splendidly under the control of Red Blanket, her beloved Indian."

RED BLANKET: "Medie-Edith's own Spirit always is outside of her physical body, but still attached to it, while I am inside her body tightly holding her."

This was our last materialization session. However, Sara continued her usual private classes, with the Great Teachers speaking through her.

THE MASTER JESUS, OUR ELDER BROTHER

My second private session with Sara was on January 14, 1957. To my bewildered surprise, the MASTER JESUS spoke to me, using Sara as his vocal channel of communication. He said: "I have established this little center from which the Master Teachers can demonstrate our work on the earth. It is necessary that many other centers also be established all over the earth." At that time I could not understand WHAT was taking place.

Again, on January 22, 1957, JESUS said to me: "You have been brought here so that you can record all of the messages given to you. Later, you will bring them to the earth people when the proper time arrives."

During all of the years that Sara and I continued our weekly private sessions, JESUS brought us a great number of lengthy instructive messages. They cover a wide range of various topics. I have only been the humble scribe.

JESUS fully materialized himself at different times during our materialization sessions. His first visual appearance before us was on December 27, 1957. He always was the last Great Master who appeared and stood with his arms outstretched towards us. He always was robed in a pure white garment. At times, some spirit voices would softly sing just prior to his manifestation. He spoke English, but often he also spoke Aramaic or Hebrew when He blessed us.

While standing before us he only made short statements. At different times he also repeated some statements.

His short messages are herewith quoted.

"I AM NOT GOD, THE ALMIGHTY CREATOR. I am merely the son of man and also the son of the Creator. Ye also are sons and daughters of the ALMIGHTY CREATOR. May the Power of the HOLY LIVING CREATOR be with each one of you."

"I am your Elder Brother and all people are my brothers and sisters. I speak with Divine Love for all people."

"This is my center and my word shall go forth and teach righteousness."

"You are the sons and daughters of the "GREAT OVER-SOUL, THE OMNIPOTENT, THE ALL-POWERFUL FATHER-CREATOR."

"Lo and behold, I am always with thee. The FATHER in I, and I in the FATHER and the Father in all of you. The FATHER was before I".

"Let not your hearts be troubled, but believe in the FATHER and yourself."

"I and the FATHER are One and I greet you in the name of the FATHER. The vibration of the FATHER's Love is working through me."

"May you walk with the LIVING FATHER-CREATOR. You are in the FATHER and the FATHER is in you. I am only His son."

"The FATHER loves me and He loves you and I Love you. That Love lasts forever."

"In the name of our HEAVENLY FATHER I bless you."

"I walk in your heart and I think in your mind. As I voice in your language so do I voice in my own language."

"I am not the 'LORD OF ALL LIFE' I am only the son of the FATHER."

"My brethren, do not violate the highest commandment. Therefore, I say unto you: 'Love ye one another, as I have loved you.' "

"I bring you the gifts of the FATHER and I bestow upon you, my disciples, the Power of the FATHER."

"Ye are my sheep and I am your shepherd, therefore, you shall never want."

"May the FATHER dwell in you as He does in me."

"May the FATHER watch over you as He watches over the dove of peace."

"If you will knock upon the door of your own temple, I shall admit you. If you will open the door of your heart, I will enter therein." (We heard the ringing of spirit temple bells just before Jesus appeared).

"Bow your head in humility and raise your heart in glory, for the FATHER is within you.—HE IS YOUR LIFE."

"You shall always see me for I shall always abide in you and you shall abide in the FATHER."

"I point the way in the garden of spirituality."

"I am with you always, my brothers and sisters."

"If you love me, abide in me."

"Touch me not and fear me not for I live and move and have my Being within you. Love me even as I love you."

"Be ye humble and be ye merciful, that you may obtain mercy."

"Love your enemies for they will dissolve under the passion of Divine Love."

"Blest is he who does not condemn, but wise is he who forgives all people."

"Fear not to tread where the Master Teachers have trod. Go forth upon your mission of mercy. Forgive all people for they know not what they do."

"Blessed be the tie that binds all human hearts together."

"Let not your heart be troubled and be not afraid. You have nothing to fear for I am always with you."

"The FATHER has sent me to help you overcome your burdens."

"Open your eyes, my earth people, to all the beauties around you. May you walk in beauty and appreciate everything the FATHER has created."

"Let your light so shine while you live upon the earth plane that you shall always live upon the level and depart on the square."

"Not a sound of a hammer or a nail was heard in the building of Solomon's temple, because it was the "Temple of the Holy of Holies, the Inner Sanctum."

"I come to greet you people. You are my brothers and sisters of the earthly flesh. In the name of our HEAVENLY FATHER I bless you." (Red Blanket, the lovable Indian who holds the medium, Edith, said: "Master Jesus, bless me, old

Indian too." (6-3-58)

"Be not afraid to bow down before the altar of truth, as the truth shall make you free. I bless each of you." (Red Blanket: "Me say, Jesus also bless old Red Blanket too, and that be nice." (9-25-58)

"My brothers and sisters go forth and express the commandments. If you love me then you love the truth and speak my language. Share your spiritual light, for as you share the light, more light will come to you. So live your life upon this mundane earth world so when the time comes you will be able to stand upon the hill and cast forth your light."

JESUS stood before us on 2-6-59. He said: "I come as you see me in the picture hanging on the yonder wall in our class room." Then he spoke to Sara: "You may approach me, sister." Sara walked over to Jesus and stood close to him, within a few inches from his face. Then again he said:

"Blessed ones of the earth, fear no evil. I walk with you and I talk with you, I am just thy Elder Brother."

"LOVE IS YOUR KEY-WORD . . . LOVE IS YOUR POWER . . . LOVE IS YOUR LIFE."

(He had shoulder-length black hair and a wide black chin beard about four inches long, looking exactly like the picture. That reproduction picture is a copy of the original painting by Gabriel Max, an Austrian who was born in 1840. Sara gave me that picture).

JESUS appeared on 8-28-59, saying: "Whither you go, know that I am with thee always. I bring you glad tidings this holy night, that each of you are baptized in the 'River of Life.' Close your eyes, each and every one of you, that I may place the 'Spital of Spirit Healing Oil' upon you. Can you feel the moisture on your eyelids?" Group replied: "Yes, we feel it." JESUS: "Know, that the FATHER-CREATOR heals through me. I in the FATHER and the FATHER in me."

DR. WEST: "He magnetically placed the oil of the GREAT SPIRIT upon your eyelids. That was a wonderful feat."

On 11-27-59, JESUS appeared and said: "I now give you the Holy Communion. Bow your heads that you may feel the inflow of life going through your body. I now bestow upon you the ointment and the oil of the CREATOR's Love, that you may reflect His Glory in the temples of time. My Peace I leave with you."

DR. WEST: "Did you people feel the 'thrill' of His Presence?" Group: "Yes, we felt it."

RED BLANKET, Edith's lovable, humble Indian, who tightly holds her during each materialization session, said: "Me shiver with shakes too. Me going to be much better Spirit."

JESUS on 12-18-59 fully materialized himself and said:

"MY FATHER has sent me to you this night. I call upon my legions of Spirit Hosts to bless this home (temple) and I bless it with the Power of the FATHER: 'Oh, FATHER, bless these thy children.' Silent night, holy night! 'May the reflection of the FATHER ever beam on your countenances. Glorify your FATHER-CREATOR Who is in heaven, but do not glorify me.' "

"I want all of you people to cleanse your mind of all impurities and come and sit at the table with me. This represents the 'Last Supper.'

"Close your eyes and you will taste the salt of this wafer, it represents my physical body. Drink of this pure wine as it represents my Life. The communion table is set for thee at all times, and this communion has been given to you. This communion means that I want to be in you and that I Love you. I want you to be in me. I am in you and you are in me.

"I am merely the 'CHRIST.' I became 'Christed Heir' long before I was born on the earth and I was a 'LIVING CHRIST' while on the earth."

"PARTAKE OF THE LIVING CHRIST, HE WHO I AM."

On 3-6-59, JESUS said: "I have been sent upon the earth and I will be seen in the busy marts of men. People will see me and they will know me. I have come in human form to the earth in this Golden Age."

JESUS, on 6-26-59 said: "I left this earth world and I have again returned. I walk in the hearts of millions of people and I will walk in the aisles of all synagogues and all churches. I have come to the multitudes and I am here before you tonight."

SARA: "He has told us in our classes that he has returned

to the earth and that he has taken up 'His Rule and His Reign' over the earth. Tonight, he has verified that information."

JESUS, on 12-5-58, appeared before us, with his arms outstretched towards us, and he was very beautiful. He said:

"Blessed art thou among the peoples of the earth.

"THOU CAN SEE ME, THE 'LIVING CHRIST' OF ALL THE AGES."

On 4-29-60 JESUS appeared before us, wearing his white robe, covered with a mantle (cape) that was all white, which had many folds. He said:

"I now heal you and place before you my 'Mantle of Healing.' " Then, with both of his hands he swung the mantle out in front of him towards all of us. It was a large mantle that had a great number of folds in it. Parts of his mantle and also parts of his robe remained lying on the floor for several minutes after he disappeared. All of us were awed at this very wonderful healing demonstration and we strongly felt His Presence.

JESUS, on May 27, 1960, appeared and said:

"I have prepared a place for you that you may not have to wait until you discard the physical flesh body which is so weak. You shall now meet me as I greet you.

"Behold, I lay before you a 'Magic Carpet' upon which you can tred with your spiritual feet as you journey on the road to Godhood."

(While he was speaking he spread out before us a great sheath of ectoplasm that extended from his hands as he spread the "Magic Carpet" on the floor before us. It covered a large area on the floor. That was a most marvelous demonstration).

May 27, 1960 was our last materialization session.

(JESUS fully materialized himself before us 25 times.

ANCIENT PHOTO OF JESUS PAINTED IN YEAR 1840. During one of Jesus' materializations, when he appeared he referred to this photo which was framed and hanging on a wall in our session room in Sara's home. He looked like this photo. —Laura Horst.

This caption is placed below the ancient photograph:
IF YOU WATCH THE EYES CLOSELY, THEY WILL SUDDENLY OPEN! 14 colors were used to get this remarkable effect.

This picture is also known as "St. Veronica's Handkerchief" and is a copy of the celebrated painting by Gabriel Max, an Austrian who was born in 1840. The original painting, valued at $25,000.00, is in a private collection in Prague, Austria. His best known work in America is "The Last Token" in the Metropolitan Museum of Art, New York.

TES, THE GREAT MASTER SPIRIT HEALER FROM A FAR-OFF WORLD

The first time on May 3, 1957, TES, the Great Spirit Healer, came to our center. He told us that he had come from a far-off world, named the "Planet of Love" which is one of the Golden Pyramid Worlds, that lie far beyond the groups of Etherean Worlds. After this TES attended every one of our regular weekly sessions conducted in Sara's home by Master Spirit Teachers. They used Sara as their communication channel. TES came to perform a Great Healing Work for people on the earth.

Before the final ending of each session, TES and his group of Associate Spirit Helpers would bring into the session room many thousands of confused, chaotic Spirit Beings, who did not know what had happened to them at the time of their so-called earth-death. They had been rescued by many groups of righteous Spirit Missionaries. Those rescue groups also would capture and bring into the session room, thousands of evil Spirit Beings, who were found roaming about on the earth. Those evil Spirits were possessing and inflicting their wicked deeds upon earth people. Also, at times, the rescue groups would bring a great number of evil Spirit Beings, whom they had captured in the lower astral heavens of the earth. The Great Masters had built highly magnetized vibrations in the session room. When those rescued Spirits were brought into those vibrations, it greatly helped to release those Spirits from the strong, pulling, earthly vibrations. TES would give them healings and then they were taken to their proper places where Spirit Teachers taught them the righteous ways of spirit life.

TES, on 3-7-58, said: "I now have 150,000 Spirit Doctors

working with me. They are going into all of the nations upon the earth, performing healings upon earth people, who have called the healings 'Miracles.'

On 5-30-58, TES gave the following prediction: "There will come a time, in the future, when there will be 'Instantaneous Healings' performed upon earth people."

TES, on 7-20-62, said: "Now, I am using different healing machines that are already built in the worlds from which I came. Eventually, those same kinds of machines will be built in your earth. They will be something like your X-Ray machines and they will project different healing colors."

This, now, explains how Sara received the painted picture of TES. It was given to her by her mortal artist lady friend named "Ionia" who was critically ill with cancer. Sara requested TES to go to Ionia's home and try to give her healing aid. Ionia had attained a degree of psychic sight so she very plainly saw TES as he stood beside her bed. She was so seriously ill but she managed to paint the original 14" x 24" picture of TES (page 71).

On April 18, 1958, in the afternoon before our session started, I drove Sara to Ionia's nearby home to pick up that painted picture of TES. Then Sara hung his picture in our session room, where it remained during all the many years as our sessions continued. (Later, Sara gave that picture to me. It is an exact photographed copy of the original painting but it is reduced in size).

On that same afternoon of 4-18-58, TES said: "That picture painted by Ionia, is a 'Living Image' of myself. If I had that picture in the spirit world, I would stand there 'alive' just as I am now standing here in this room. In the spirit world EVERYTHING HAS LIFE AND NOT ONE THING STANDS STILL. I say to you Lara, visualize me as standing here and that I look the same as you see me appearing on the picture." (Sara always could plainly see TES and the other Great Teachers who spoke through her).

TES fully materialized himself on 4-25-58, during a materialization session. He said: "I think that picture of me is a good likeness of myself." During the several other times when TES materialized himself before us, he always wore the same light-blue toga-style garment, and a pointed conical headdress, denoting a symbolical identification.

On 5-3-58, during my private session with Sara, she described TES: "He looks and is dressed exactly like his picture. He is standing here with his eyes open. They are very large and very dark and they have a kindly expression that I cannot fathom."

IONIA died November 26, 1958, and her funeral was to be conducted on November 30th. During a private session prior to her funeral, Sara and I were astonished at what took place.

On November 28, two days before Ionia's funeral, a Great Master named MATHENO was speaking through Sara. Then, suddenly, I became very startled when Sara began to loudly shout, "OH! OH! OH!" because she was experiencing what she later said, was a strong shock of electricity. However, that shock was not originating from Matheno. It came from someone else who was too excited to await her turn to speak, when the vibration would have been lowered by the Teachers. A lower vibration would have prevented the sudden electric shock that Sara experienced when that other Spirit Being took hold of her.

In a very loud, highly excited voice, the intruder spoke through Sara, saying: "I am Ionia and I am glad that a man brought me here. I want to tell you I saw my husband at the time I was dying and when I saw him I was so frightened because I had thought he was dead, but as soon as I stepped out of my flesh body there he was! Oh, I cannot tell you what a shock it was to me when I saw him! Oh, bear with me and help me, because I need help. I was so stubborn in my mind and I did not want to know the things that I know now, but I don't know much yet about this spirit life. Oh Sara, you are the only one who always tried to help me. You will never know what this contact means to me." Then Sara said to Ionia: "I have always loved you and wanted to help you. Now, Ionia, I have asked ADONTUS and PO and Others, to take care of you." Ionia replied: "I know you asked them to take care of me because they were there to greet me. Oh, everything is so blurred and in such a great confusion. I know that TES can bring me out of this condition."

Then PO said: "You know that Ionia's physical body is not yet buried. We don't know now if we will allow her to see the burial of her flesh body, but we will be there with Sara.

"Now you see dear Lara, what the 'passing' means. It is the transition from one state of existence to another state of existence. We were at Ionia's bedside during her transition."

Then TES spoke: "We are today taking the healing power to Ionia. Now that she has shuffled-off her physical body, we are going to heal her spirit-body, because there is still much to be done."

Matheno again continued speaking: "I wish to make a statement that I came in here this afternoon prior to this time to bring you a message which was so suddenly interrupted. I was aware of the fact some of the vibration here was today to be used to help a Soul (Ionia) into the spiritual light, but that information was unknown to Sara and to yourself, Lara. Now I will continue with my message."

After Matheno finished speaking then other Teachers brought their messages. When our lengthy session was concluded, Sara said: "I say to you, Lara, that electric shock I received when Ionia grabbed hold of me was an unexpected experience as it nearly knocked me off my chair. You can imagine me as I was listening to Matheno speak; I seemed to be half-asleep and half-awake and I was just talking, saying one word after the other. Then all of a sudden Ionia took hold of me. It seems that when the Spirit Beings speak through me they have to get inside of me, or get into my electrical aura surrounding me. Ionia's slipping in was such an electric shock to me. That really was something so unusual that I experienced."

I replied to Sara: "It frightened me as I could not imagine what was happening to you, but I knew your Teachers were protecting you. What happened today was a most startling experience I shall never forget."

After TES made his first contact with us he attended every class session that was conducted in Sara's home. He also diligently performed his healing work with people throughout the earth world.

The first time that TES fully materialized himself before us was on 12-27-57. We were so happily surprised when he dashed out from behind the black cotton curtains that composed the materialization cabinet. He quickly walked out into the center of the room holding his arms and hands up high, and with great force and power he said: "I AM TES. I want each of you to come up to see me. This is the first time I am appearing in this way before you."

SARA said: "TES, we all want to thank you for all of the healing work you have done for us." TES replied: "Do not thank me, thank the CREATOR, Who has done it all. We Spirit Masters claim no credit because the CREATOR works through mortals and also through Spirits HIS Wonders to perform. Remember, all you have to do first is 'Heal Thyself.' And when man heals himself then he 'knows' himself. So fear not as the CREATOR is always within you. There is healing for the Soul of man and healing for man's physical body. Know that everything is possible. You are Spirit Beings now and you must learn the power of constructive thoughts as they are thoughts of health. It is destructive thoughts that cause diseases. I say to you, Sara, I work through your hands when we do healing work for mortals who are brought into our classroom. I also work through your hands when you remove Spirit entities from mortals who come to you for help. We both are channels for that work." Then TES quickly disappeared from our sight. His conical headdress was made of gold and there were beautiful jewels down the sides of his light-blue toga-style garment.

TES appeared on 1-31-58, saying: "I want to touch all of you people in this room with the Divine Healing Power. I want to touch each one of you with my holy garment that you may realize I am here before you." Then he walked over to each of us and touched us with his garment, then he quickly disappeared.

On 2-28-58 TES appeared, standing before us. He said: "I want you to know that the God who rules the Crystal Worlds has taken the healing power from this center over all of the earth world. I want each one of you to realize that you are a dynamo of power unto yourself and that you can heal one another."

On 5-30-58, TES appeared, saying: "I want you to know that I am here at every session. You people can also heal anyone in Europe as distance is no barrier. Because the flow and flux of power and the flow of thoughts cannot be stopped by distance. You can heal in foreign countries while you remain right here. That is what I want you to do. Send out your healing thoughts and change the atmosphere."

TES appeared on 6-13-58. he said: "Each one of you have already received a healing this night. May the CREATOR'S Power flow through all of your bodies and may each one of you feel the waves and the currents of that divine energy and that vital force that flows from the universe to you. This is but a kernel, a seed, that has been planted here and out of it will grow a great work. The physical, mental and spiritual healing must go on."

On 7-25-58, TES appeared saying: "Blessed ones of the earth, and especially to you Sara, my beloved channel, greater work has not yet been brought forth, but you are doing a great work. In the future there will be miraculous (so-called) cures for the people of the earth." Sara replied: "Thank you TES." Then TES said in a very loud voice: "Do not thank me, my dear, but thank the CREATOR, as we are only the messengers. There are many Golden Age Masters here tonight."

TES on 2-6-59 appeared, saying: "I AM TES." Sara said: "Oh, TES, we know how much you have helped us." TES replied: "It is not I, it is the CREATOR'S Power that works through me and I work through you."

On 6-26-59 TES appeared, saying: "Good evening my friends and my beloved channel. Come forth my dear Sara." As Sara stood in front of TES she said: "Dear TES, you do such wonderful healing work for us." TES again, as always, replied: "No, it is the CREATOR who does it. I am the instrument for the CREATOR and you are the instrument for me." Then TES gave Sara a healing treatment with his fingers. Then he called out the names of each one of the group to come forth and stand before him while he gave each one a healing treatment. I was the last one to receive my treatment. That was an astonishing experience for me.

Just before TES gave me my treatment he excused himself and quickly disappeared behind the cabinet curtains. Then again, quickly, he reappeared and said: "Come forth my dear Lara I have returned with a special 'magnetic blanket' that I am going to place around your shoulders. Turn your back to me and fold your arms across your breast, as I want you to see the spreading of this wonderful blanket as I put this blanket around you." I replied: "Thank you TES. I do feel the blanket but it is very hot." TES said: "Of course it is warm as it is a magnetic blanket. Now close your eyes, my child, and it shall go into your body. Nothing like it has ever entered your body before this time. Now your nerves shall be made to rest easily under the influence of this blanket. It is a special blanket just for you." Then he quickly unwrapped the blanket from around my shoulders and my body, saying: "Go forth, my child, to your chair."

Then TES spoke to all of us in the room, saying: "My friends, this is the blanket that I also use to treat those people who have infectious diseases. And now, I spread this blanket out for all of you to see it." We very plainly saw the spirit blanket and estimated that it was about six feet square. Then TES quickly de-materialized the blanket by giving it a quick swish and it disappeared. Then TES also disappeared from our view.

DR. WEST said: "When TES came back behind the curtain to materialize that blanket, I watched him. With the snap of his fingers, that were quicker than the flash of an eyelid. He manufactured that magnetic blanket and then quickly brought it out to use on Lara. That was an outstanding demonstration and I could feel the electricity 'snap.' "

TES appeared on 9-25-59 and said: I am here most of the time, but I am a Universal Worker and I go everywhere. I will not be possessed by any one individual person only by the CREATOR. And I don't want you people to ever be possessed by each other. Be universal in your mind and be universal in your own Spirit. Sara, can you see my garment?" She replied: "Yes, with all the glitterings upon it. You are very tall." Then TES disappeared.

On 1-29-60, TES materialized before us and it was his last time we visually saw him. He said: "I inspire each of you to help others as there is much work to be done here on the earth. I notice that you still have my picture on the wall of the session room and it looks like me. Sara, do you see my headdress that I wear tonight?" Sara replied: "Yes, I see it and it looks just like your picture." Then TES said: "I would like to show myself to all of you people in the room so that you can plainly see me. I want every one to see me as I stand here in front of you." All saw him before he disappeared.

(TES fully materialized himself 10 times)

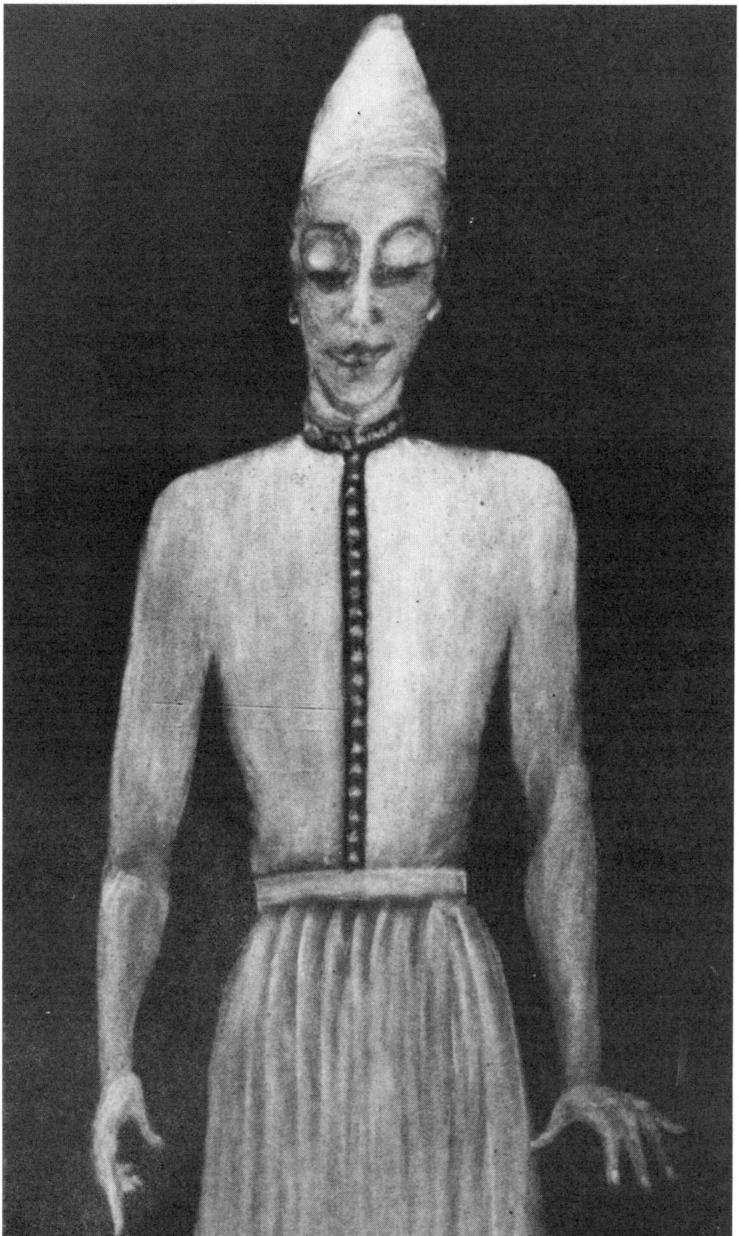

TES . . . THE GREAT MASTER SPIRIT HEALER **who came from the far-off solar system of worlds. He did great healing work during the many years he worked with us at Sarah's center.**

PART 3

GREAT MASTER SPIRIT BEINGS WHO BROUGHT MESSAGES

MOON GOD MESSAGES
MESSAGES FROM MARS
MESSAGES FROM VENUS
MESSAGES FROM THE PLEIADES
MESSAGES FROM THE PLANET URANUS
JUPITER, SATURN & PLUTO
MESSAGES FROM THE STAR BEINGS
WORLD OF ESMERALDA
RUBEON PLANET
OTHER FAR-OFF WORLDS AND THEIR PEOPLE
LOVE WORLDS
PYRAMID WORLDS
A FAR-OFF SPIRIT WORLD
MIZPAH ORGANIZATION
SEVEN GOLDEN AGE MASTERS
GREAT BEINGS BROUGHT MESSAGES
SPHERICAL . . . A MIGHTY MASTER
GOLDEN CHARIOT BEINGS (UFO'S . . .
RIDERS IN THE SKY)
SEVEN GREAT FEMALE BEINGS
PURPLE RAY BEINGS
SEVEN RAY BEINGS
MORFIA BEINGS BRING MESSAGES

MOON GOD MESSAGES

During our afternoon session of December 27, 1957, a Spirit Being named GOD OF THE MOON made his first contact at our center. He said he was living on a far-off Pyramid World when he was appointed as the CHIEF GOD to rule over the earth's Moon. We had a materialization session that same night and he appeared before us, saying: "I live on your Moon, it is my abiding place. That is why I come arrayed in blue. There will be other visitors coming from the outer spaces in the days to come." He was of tall stature.

He also made several other contacts with us, bringing information about the Moon. Herewith are a few important excerpts from his messages.

"The Moon has a civilization to such an extent that the people now living there are very far advanced spiritually. They are farther advanced than earth people. They are man-type Beings, such as earth people, but they are very tall in stature. They live in a different dimension of life, in a different rate of vibration that is not the same vibrational rate as operates on the earth. They live in a vibration of Peace, Love and Joy, and there is no hatred, greed or selfishness, because they have developed their spiritual attributes, such as Brotherly Love, and they practice those talents."

"The Moon is inhabited by millions of highly evolved Spirit Beings who function through their spirit bodies. However, they can also create for themselves, when they so desire, an especially created simulated earth-type of a body that enables them, at times, to function in materiality and also in spirituality. Now you may say: 'That is fantastic.'

But I say to you, 'That is not really fantastic. It only seems so because earth people do not understand the operation of the higher spiritual law of life.' "

"We have come to the earth people as helpers and as friends and for a definite purpose. The earth is widely known throughout the universes as the 'Red Star, Bloody Planet' due to its many bloody wars and the unrighteous ways of living demonstrated by a great number of its people. The time now has arrived when all earth people must live together in spiritual brotherhood and love.' We the people who inhabit the Moon in our spirit forms, can accomplish many more things than earth people can accomplish."

"The earth and its heavens are to become blended together in harmonious operation during this New Spiritual Golden Age. Therefore, all things must change upon the earth and in its lower heavens and all things must be made new. What affects the earth also affects the Moon and all of the other planets in this solar system are vitally concerned about the unrighteous happenings on the earth. They are coming to the earth, working together as one unit, helping to bring earth people into righteous ways of living."

"The earth is undergoing a thorough cleansing, which is being done by unusual holocausts of different kinds. This cleansing is necessary to rid the earth from its evil vibrations and its iniquities, crimes and wrongdoings performed by the unrighteous people. The earth's lower heavens also are being cleansed of the evil Immortal Spirits, who roam about on the earth's surface, inspiring mortals to do many wrong things, such as crimes and arson. This thorough cleansing is necessary to prepare the earth and its people so they can live righteously together."

"We have been giving our instructions and are marshalling our numerous Hosts of Beings to do a great work for the earth. I am the MOON GOD and I am in command of a fleet of space ships that are watching what is taking place upon the earth. Our space ships are guarding and protecting the earth, to offset certain things of a destructive nature, so that such things that are not to be destroyed will be preserved. Again, I say to you THERE ARE GREAT FLEETS OF

SPACE SHIPS THAT ARE PATROLLING THE EARTH. Do not fear, as the earth will NOT be completely destroyed during its cleansing."

On January 15, 1958, the MOON GOD said: "A large number of Great Masters recently have built and are placing a magnetic belt around the earth. It goes far up into the atmosphere and far out into the universe for many miles. It is now in operation and is used to help protect the earth. The time will come when the earth people will become fortified in that magnetic belt."

"Mortal earth people will be able to conquer outer space and travel to the Moon and to other planets and worlds. They will be able to do this through the aid of this magnetic belt. That magnetic belt will be the means to help your airplanes and space ships to be propelled to those other worlds. Those things will be done by using the belt and the 'free energy' existing in space."

Mortal earth men will be able to go to those other worlds, because this body will have become properly magnetized, so that he will be able to breathe and live in other worlds. This magnetized procedure will take a long time. But the individual persons probably will not be aware of what is happening to them. Then, when mortal man leaves the earth to go to the different worlds he will be able to see what is on those other worlds and be able to bring back information about those worlds.

Most earth people will be unable to understand all of the many unusual happenings taking place on the earth during this New Age.

February 14, 1958, the MOON GOD gave the following prediction:

"THE SPACE SCIENTISTS IN THE UNITED STATES WILL SEND TWO MEN TO THE MOON, HOWEVER AFAR OFF THAT TIME WILL BE, AND WHEN THAT TIME COMES, TWO MEN WILL LAND ON THE MOON IN THEIR SPACECRAFT."

"Your space scientists are working very closely together. Also, a large number of High Raised Immortal Spirit Beings who are living on the Moon, are inspiring and helping your space scientists to bring space flights into fruition. The Great ones from the Moon *have come to the earth,* and they are impressing your scientists what to do. Those from our Moon World are the ONES that will make it possible for this moon trip to be successful. That is the way the ALMIGHTY CREATOR'S Plans work out. HE sends the more highly advanced Spirit Beings to the earth to work with the earth people. Throughout all the ages of eternal life, the more highly evolved Beings constantly work to help the lower Ones. This Moon trip is not something that is engineered and brought into manifestation just through earth people. IT IS TIED IN WITH THOSE BEINGS FROM THE MOON. THOSE GREAT MASTERS FROM THE MOON ARE WORKING WITH MORTALS ON THE EARTH. THAT IS THE WAY IT IS BEING DONE. Yet, mortal man and especially the scientists, know nothing at all about how the Moon people are inspiring them what to do."

THE MOON GOD'S PREDICTION CAME TRUE,
JULY 20, 1969.

At Cape Kennedy, Florida, on July 16, 1969, the APOLLO II spacecraft was blasted-off by a Saturn V rocket. Aboard were THREE MEN, Neil Armstrong, Edwin Aldrin and Michael Collins. On July 20, 1969, Neil Armstrong was the first man to land on the MOON and Edwin Aldrin was the second man who walked on the MOON. Michael Collins commanded the spacecraft and encircled the MOON during the time while the other two men walked on the Moon and gathered up rocks. Then later, Collins picked up Armstrong and Aldrin and all three men returned safely to the earth on July 24, 1969. THIS EVENT AMAZED ALL PEOPLE ON THE EARTH.

On March 28, 1958, the MOON GOD materialized and said: "Be not dismayed that there will be many strange visitors to your earth. There will be those who will come in space ships and they have been seen. There will be demonstrations and signs in the sky. There will be a great phenomena all over the earth world."

Again, on April 25, 1958 the MOON GOD appeared before us. He said: "I am a Universal Worker because there is no partiality as all of you are 'Souls' migrating along the highway of life. We occupy the various planets in your solar system. All planets are inhabited except the Sun of your solar system. The GRAND CENTRAL SUN feeds your Sun. The Central Sun itself is filled with organic life only. That organic life is what feeds, takes care of and warms your earth. You could not live without that Sun. You also need all of us who come from the various planets." Then Sara said to the Moon God: "Are there people living on the moon?" He replied: "Yes, absolutely and that is my abiding place when I am away from the earth. That is why they call me the GOD OF THE MOON because I have authority over there and over many 'Minds' that I have helped. And in truth, Other Beings have authority over my own Mind. I have my Mercy Emblem on tonight on my head."

I then asked, "Are there any mortal Human Beings on the Moon?" He replied: "There are no human flesh bodies. There is spirit life on the Moon in our kind of spirit body. I am very tall tonight. We are a race of people who have peopled the earth eons ago. Formerly, I was one of the ancient Atlanteans. All of you were Atlanteans."

MOON GOD appeared on June 13, 1958, saying: "The MOON is a great world unto itself. *IT IS INHABITED.* Many Gods from the friendly skies are visiting this center."

On November 7, 1958, the MOON GOD appeared and said: "You are living in a spirit circle right now because the atmosphere is a spirit circle. Your Sun warms the earth and all of the other planets in your solar system, as they go through their own orbit and vortex of electrical power, whirling through space. The cosmic rays of healing are permeating through the atmosphere and the nearer the light rays come the MINDS of earth people are becoming evaluated. Earth mortals will have visitors from the various planets, such as MARS, VENUS and SATURN with its many rings. Some of those visitors will be vaporous and they will be mingling in the crowds of mortals. They will appear and then quickly disappear. Earth people must be *conditioned* to see them, otherwise they would lose their sanity."

On June 26, 1959, the MOON GOD appeared, wearing a beautiful glistening robe decorated with jewels. He said to my husband, Tony; "I look upon you as a Spirit Being, as myself. Your Soul is growing and that is all important."

On September 25, 1959, the MOON GOD appeared, saying: "I have brought you a radiation on my jewel cap that I wear. It has the lustre and the blue of the Moon." Then he bowed down his head so that all the people in the room could see his magnificent headdress, radiating many colors."

November 27, 1959, the MOON GOD appeared: "I am the MOON GOD and I come to the earth to help the earth people to rehabilitate themselves with our Moon World. You will have other visitations from the Moon. The Moon is inhabited and it is a very prosperous planet.

"Space is an intervening matter between all of the worlds.

"The earth upon which you are migrating is called the 'Hell Planet, or the Purgatory Planet' of this solar system. The earth is the only material planet that is like that. But it can become a heavenly planet. But it can not as long as the minds of mortal people are full of greed, selfishness and vice."

Sara asked the MOON GOD as he stood before us on 11-27-59, "Will earth people ever be able to have a space machine with which they can reach out into other worlds?"

MOON GOD replied: "We already have them on the Moon. But your earth people can not now have anything like that, which equals the mechanism of the human brain and the human mind, because it would be dynamite in the hands of some of the evil earth people at this time. On the Moon people from the different planets are migrating. We do communicate by the power of 'Thoughts.'

"WE DO HAVE DELICATE MACHINERY THAT WE CAN USE TO REACH OUT INTO THE OUTER SPACES FROM THE STRATOSPHERES OF OUR WORLD. Those things cannot now be given to earth people as they would be wrongly used."

May 27, 1960 the MOON GOD appeared: "I greet each one of you and I bring to all of you the power that you need. We, the Masters, are becoming more as Universal Workers. We form the great battery and work together through the humans on the earth. It is vitally necessary that we do this. There will be catastrophes and calamaties upon your earth. Recognize certain facts as they are, but do not let them worry you."

(Other Spirit Beings have told us that it is possible to inhabit a world "within it" in the same way as a world is inhabited on its surface).

(The MOON GOD materialized himself before us thirteen different times).

MESSAGES FROM MARS

On September 1, 1958, THE STAR BEING from the Pleiades spoke through Sara and gave us this message about the Planet MARS.

"The Friendly BEINGS who are coming to the earth from MARS, are coming to help people on the earth. They are here to make war on the Evil Invisible Beings who inspire earth people to wage wars. The Mars people divert wars from other worlds, such as they are now doing on the earth. That is why Mars is called the warring planet."

On September 25, 1959, MIDGET MAN, DR. MARS, from the Planet MARS, made his first contact at our center when he fully materialized before us. He said: "Hello everybody! I am very happy tonight to come here from Mars . . . that is my home. I am not tall like you people of America, but I can still come and give you much power and give you knowledge about MARS." Sara said to him: "Are your space ships here?" He replied: "Our space ships are getting ready to land! They are taking some measurements of the space in the desert and on the high mountain tops where they will eventually land."

DR. WEST said: "He was only three feet in height. Some of the Martians are very small in stature, but they have a large brain. He is a Master Teacher and has a man's voice."

October 14, 1959 DR. MARS again appeared, saying: "I come from Mars. I may be little but I am big in mental stature and size makes no difference. I say to you, there will be wonderful revelations that will be brought to you here at this Golden Age Center of Light. We do not know time as all things are brought about on *vibrational vibrations.*"

On October 23, 1959, DR. MARS spoke through Sara: "I am here on the earth to help make over the earth world. Earth people don't know very much about life. All of these Great Master Teachers who come here—they know so much more than I know, yet I know a whole lot more than earth people know."

On October 30, 1959, DR. MARS materialized and said: "I am happy to greet all of you people. There are many events coming to the earth. You will be surprised at the many manifestations in the days to come."

On November 13, 1959, DR. MARS spoke through Sara, saying: "I am here, but if I had no body at all, I would still have a MIND, and I could speak to you. I am here to learn things that the Highly Advanced Teachers are bringing here to you. I want to learn the principles that are being invoked, as I want to take those principles to the people on Mars. I had not seen, and had not heard what you have been told here tonight. I appreciate hearing and seeing the different things that the Great Ones are demonstrating. I am vitally interested in those messages involving this solar system of which my world Mars is a part. I want to take that information to my world Mars. I will tell my people that I have knowledge and information from higher sources. I will tell them what has been told here tonight."

June 2, 1961, DR. MARS spoke through Sara: "I come here for a special purpose, so that I can draw from the magnetism that you people here throw off from your bodies. It is phosphorus, having power and strength and it has a blue and purple radiance. I pick that up and I take it to Mars. I use it for a certain work of development for different Beings that are on Mars. You are not aware of this work that I am doing. That is the same method that has been employed to help you people here. The Great Masters have brought to those of you at this center, magnetism and other substances from worlds in this solar system. Those substances have helped to develop the magnetism around your bodies."

On November 1, 1958, as Sara and I started our session she psychically saw a man in the room who was dressed in armor, sitting on a great stallion. She said: "I have never seen him before this time and don't know why he is here."

Then the man started speaking through Sara: "Good afternoon, my friends. I have not spoken to you before. I am the one who was here on the stallion. I am known as the LORD OF MARS and also known as the MAN OF MARS from the Planet MARS. I am here to bring great power to this center.

"The Planet Mars has been given the credit of being a warring planet and that we bring war to the earth world, which is entirely erroneous teaching. We do not bring war to the earth or to any other planet. We go to the planet where war is taking place. We go there with an invisible army and we rout the other invisible army that is fighting for inharmony, destruction and evil, so that the army which is fighting for justice and right will overcome their opponents through our help. We fight in the invisible to bring the right and the good to the planet that has war taking place, such as on the planet earth.

"That is why I am here today, to explain to you that the army of the invisibles are poised and are ready to strike against any group of warriors on the planet of the earth who are planning to start a war on the earth. It is the battle of the invisible army from Mars that is keeping them from carrying out their ill designs.

"I am bringing this information to you and to tell you that we have come here to this center to get the power that the Great Masters have built up in here. That is why I showed myself to the medium on my powerful steed. That steed symbolizes the power of my invisible army."

July 17, 1959, the LORD OF MARS said: "The color of red is a warm and powerful color. It is the leading color on Mars, and is why Mars is such a powerful planet. Again, I say to you, that MARS is NOT a warring planet. It is a planet that makes war upon the evil things that operate on your earth world. That is what MARS is doing for the earth. It is destroying the evil that is so prevalent upon the earth."

April 25, 1960, The LORD OF MARS showed himself to Sara and he wore a pointed headdress. He said: "I come to tell you that the planets in your solar system have civilizations of BEINGS that are much farther advanced than the earth people. That is, insofar that they do not have murder and wars and all of the other destructive events ever going on among the people living on those different worlds.

"MARS is not a warring planet, only to this extent, that they do not war about selfishness and the desire to destroy its inhabitants. What they do war about, is to keep the destructive negative things away from Mars, and that is where we war, so that no outside evil influences can come to our planet, such as would come from your earth world.

"I say to you, until mortal earth man can redeem himself to the point that he does not have the desire to war and murder and lust . . . until that point is reached, no mortal person will ever gain access to the planet of Mars. We are highly evolved in comparison to the mortals on the earth. We are evolved to the extent that we use the principle of Love entirely in all of our operations and in the work that we do.

"We do not operate solely through THOUGHT. We have a language which we express audibly. However, there are worlds and planets where their inhabitants do not speak audibly and do their communicating through THOUGHTS. The Planet of Mars has not yet reached that height, but Mars is far beyond the earth in its spiritual advancement. Once the earth evolves to where we are, then you are going to demonstrate the Brotherhood of Man and the New Heaven and the New Earth. The New Heaven has already reached Mars and the New Mars is now operating on Mars.

"It has taken this medium, Sara, a period of many years of her spiritual development before we, the Teachers, have been able to use her as a channel of communication as we now are using her. We are the Ones who are giving the information and you are at the receiving end. We use Sara as a telephone channel to bring through her information that earth people now are ready to receive at this time.

"I, the LORD OF MARS, say to you, this decade of the 1960's will be very outstanding. Earth people are in a NEW DAY and will do things differently and you will have satellites filling the sky.

"Not until mortal man has proven that he has been on

other planets and has brought back the 'proof' of what they have found, will earth people believe that life exists on other planets. They will find proof through pictures they will take before they get there and after they have arrived there. They will take pictures of the BEINGS living upon those other planets. They will bring back trophies and such things given to them by the natives living on those different planets. That will be 'proof' to the doubting people. Such doubters must be shown proofs, as they cannot believe anything excepting as they cognize it through their physical eyesight and vibrations of sense perceptions that they can see, hear and feel.

"Earth space scientists already have blueprints and plans and the where-with-all to build the NEW AGE SPACE PLATFORMS IN THE SKY. During the next ten years earth people will make more progress than they have made during the past one-hundred years. That is how rapidly these New Age things are moving ahead in this New Age. This NEW AGE is something you have never had before.

"Why are there so many upheavals all over the earth? Why are people fighting those people in power in politics? It is only where those in power are selfish and want to control other human Beings. Many earth people will lose their lives through fighting each other within their own country. BUT THERE WILL BE NO MAJOR WORLD WAR. Not until the right persons are in control and practice righteousness, giving people peace and dignity, will the earth begin to operate as the New Heaven and the New Earth. But it will take many years to bring this about in the many different countries and nations. To accomplish this great work, it is necessary that numerous GREAT BEINGS use the power and the magnetism that has been built up at the various centers and take it to the earth people, to inspire their minds with altruistic, righteous thoughts. This is done while people are asleep."

(This message dated 4-25-60.)

March 24, 1961, The LORD OF MARS psychically showed himself to Sara and she described him. "He is very tall and wears a metal headdress that looks like a warrior."

He said: "Mars is a very powerful planet. I appear here alone, but I have a great number of Co-Workers here with me, who are on the outside, and there are so many that they reach over a great terrain of many miles. They are Immortal Beings who have formed themselves into an army, the 'Army of Armegeddon' that is going through the atmosphere, through the astral regions, doing cleansing work over the earth from our side of life."

* * * * *

GOD OF MARS WHO IS THE CHIEF GOD RULING MARS

The GOD OF MARS had contacted our center many times and he told us his name is MARTINOLA.

On November 27, 1959, he fully materialized himself before us. He said: "I am MARTINOLA and I come from the Planet MARS. I just now came through space and I still wear my helmet."

During my session with Sara on July 8, 1960, the GOD OF MARS said: "We are bringing strength and power to the earth. All Human Beings are under the jurisdiction of the CREATOR. We are all 'ONE' and there is no separateness and no distinction. But there is a distinction of a degree of knowledge and wisdom in every BEING.

"My Planet MARS is a little farther advanced beyond the way the earth has progressed. At one time Mars was in the throes of going through the same things that the earth is experiencing at this time. The people on the earth are going through a holocaust that is reaching out into many facets of the earth. It is not coming all at one time, but it is coming a little at a time, going all over the earth."

On February 10, 1961, GOD OF MARS spoke through Sara: "I am the GOD OF MARS ruling over the world of MARS. I say to you that MARS IS INHABITED with BEINGS like yourselves. They have bodies similar to yours. But their minds are flexible and far in advance of the minds of earth people. However, there was a time when Mars was on the same grade as your earth now is. We had to work our way out and up, the same way the earth now is doing.

"We have on Mars a system, but you would call it a government. Our system is on a much higher scale or plane of understanding, because we demonstrate LOVE entirely. It is LOVE and SERVICE for each other."

On August 11, 1961, GOD OF MARS spoke through Sara: "I have been drawn here tonight through the power that the GREAT ONE displayed here this night. The power he brought to this center was like a streak of lightning and I read what it displayed. I read of what is coming to this entire solar system, of which the earth and my Planet Mars are a part. That power will change all things on the earth and also on Mars.

"I would like to shout this from the housetops and appear in the sky and warn all unrighteous people on the earth, telling of what is coming, unless they change their wrong ways of doing things. I would like to tell those people who are operating their governments in their own selfish ways of greed and power and with their secret destructive thoughts, that unless they apply love and compassion towards their brotherman, they will face a condition for which they are not prepared."

May 27, 1960, The SCARLET PLANET GOD from MARS, fully materialized himself before us. His message is quoted in the materialization chapter. He is not the God of Mars, but is a Co-Worker God working with him.

THE CHIEF . . . AND . . . THE CHIEFTAINESS FROM MARS

On June 6, 1964, Sara psychically saw TWO BEINGS. One was a woman robed in copper colors, who wore long copper earrings and she had a big headdress. The other BEING was a man. They mentally told her they came from MARS.

Then on June 10, 1964, when I attended a session with Sara, she told me that ever since she first saw those Two Beings, they have been here at the center with her. That they are the color of copper and their garments are made of metal. This night they both are here. She wears a bronze-copper metallic garment that extends out in folds, forming part of her uniform. It sticks out below her hips. It is very beautiful. She wears a very high headdress and it has corners on top of it, something like an antenna that projects way up high. Her face is the color of bronze and it does not appear to be composed of flesh like ours. Then Sara touched her and said: "I feel her hands and they are smooth metal and her face is of metal and she has very oblong eyes which are closed. But she has features like we have, though they appear to be metal. She says she came from a space ship.

"The male Being wears a uniform that appears to be all in one piece, and he wears boots. He also wears some kind of a coat that is made of metal, and a bronze helmet having a funny little shape at the top, like a weather vane.

"The TWO BEINGS are joined together by a two-inch tube that looks like a rod. It is not bronze and I don't know what it is. Now, I felt it; it is flexible, but it is not rubber. It contains a liquid that looks like quicksilver. It moves back and forth and it has bubbles in it."

Then the FEMALE BEING spoke: "My brother and I have come from the Planet MARS, COMING HERE FOR A DEFINITE PURPOSE. I am a HIGH-RAISED CHIEFTAINESS. Our work pertains to space ships. My brother is a CHIEF, and in our sphere of life, we are considered as very High Raised Beings."

THE CHIEF said: "We will play a very important part here at this center. We are not here in the flesh, but are here in the kind and type of bodies that we function through on our Planet Mars. You will hear about many different descriptions of people coming from Mars. As you will recall, the GOD OF MARS and the little midget, DR. MARS materialized themselves before you. There are many different types of Beings on the different worlds. Tonight, we show ourselves in the way we manifest and operate on Mars.

"In the future, earth people will accept the landing of space ships in the same way as they now accept the landing of airplanes. In the far distant future you will see vehicles that will be operating upon the airwaves. You can steer them and travel from place to place with ease and comfort and without accidents. THIS IS THE AIR AGE. You will see many different vehicles and things in the sky that will 'stun' you and surprise you."

MESSAGES FROM VENUS

August 22, 1958, THE VENETIAN from VENUS materialized himself before us. His message is quoted in the materialization chapter.

August 29, 1958, THE VENETIAN spoke through Sara: "I come from the Planet Venus, which represents LOVE in all of its phases. It is that love vibration and the magnetism from Venus that I am bringing here. It will instill greater love into your hearts and souls. In the distant future, the time will come when people on the earth will display LOVE in all of its GLORY."

September 5, 1958, THE VENETIAN spoke through Sara: "From my plane of existence, from Venus, LOVE reigns supreme. There is Love, Harmony and Peace, and war is not known on Venus. That same kind of existence is what is coming to the earth. But first a thorough cleansing of the earth must take place."

November 21, 1958, The GOD OF VENUS was brought into our classroom by the MOON GOD, who introduced him to us. Then the GOD OF VENUS spoke through Sara: "I am glad for the opportunity to speak and shall attend every future session at this center. There are many billions of worlds that make up the GREAT WHOLE OF CREATION. The entire CREATION is so fantastic, so colossal, so vast, your own imagination is not able to grasp it."

July 31, 1959, GOD OF VENUS materialized before us. His message is included in the materialization chapter.

ORLANDO, A GREAT MASTER TEACHER FROM VENUS materialized before us on January 2, 1959, July 31, 1959 and on April 20, 1960. His messages are quoted in the materialization chapter.

November 29, 1962, THE VENETIAN spoke through Sara: "I am a Great Teacher from Venus, and you, Lara, will write about Venus."

ISHMAEL . . . HE CAME FROM VENUS

May 15, 1958, A Highly Evolved Spirit Being came to our center, bringing with him a Great Host of Far-Advanced Beings, to help him do his assigned work on the earth. Speaking through Sara, he spelled out his name: "I am I-S-H-M-A-E-L, ISHMAEL. I have lived on many different far-off worlds and planets, but at this time I come from the Planet VENUS. I have always helped with the raising-up of the younger worlds, such as the earth. That is why I now have come to the earth to help raise up all earth people. The time is now here when more farther advanced concepts and new things are to be displayed and manifested upon the earth.

VENUS is called the Planet of Love and it has a golden magnetic belt surrounding it. VENUS has a population of very beautiful, perfectly formed Beings. They are built somewhat like the forms of earth people, but they are a thousand times more beautiful. Their perfect bodily forms reflect their highly advanced spiritual development. Their spiritual evolvement is reflected through their beautiful auras surrounding them. Their spiritual auras extend to about two-hundred feet out from their bodies, and their beautiful auric luminosity is beyond description.

May 8, 1959, ISHMAEL again came to us, bringing with him another GREAT BEING. Sara psychically saw both of them in the room. She briefly described ISHMAEL. "He has a black beard and he wears a robe that has a lot of blue and white in it." Then she described the other GREAT BEING. He wears an odd tall headpiece. It curves in and out in what looks like scrollwork. It has a big gold decoration that is about one foot high that goes up in the front of the headpiece, and it is in curves. He also wears a band around his head. It has points.

Sara further said: "As I look at him, he has THREE EYES. His two regular eyes are half-closed, but the EYE IN THE

CENTER OF HIS FOREHEAD IS NOT LIKE HIS OTHER EYES. IT PROTRUDES WAY OUT, AND THAT EYE SEEMS TO SEE EVERYTHING. I have never seen this BEING before this time."

Sara continued describing the GREAT BEING: "Now I see he holds a short spear with a spearhead, and he points it towards you, Lara. He wears a very peculiar robe and it appears to be white. The bottom of the robe has broad gold scallops and at the end of the scallops there is a gold fringe that is about two inches wide. Around his waist he wears a gold girdle. It is crushed, with its ends also scalloped with the gold fringe. Down the front of his garment, from the top to his waist, are gold knobs which we might call buttons. Around his throat and neck there is a gold band that stands up like a collar. He mentally tells me that his garment represents his spiritual advancement and the high office that he now holds."

Then, the GREAT BEING said: "My name is ALLAH, and I say to you, each Master Teacher who comes to this center, has a certain power and a certain intellectual message that he brings to you. My THIRD EYE has evolved to a great extent and it is very prominent in the middle of his forehead. Its purpose is for the use of spiritual sight, for intellect and for obtaining knowledge. I am not in a flesh-body but in a different kind of a body, I come from many worlds.

"The heavy flesh body of a mortal earth person usually retards the spiritual growth of his own personal Spirit. The flesh-body many times endures great pain and sufferings, which are a part of earth life experiences. However, at the time of so-called earth life death, when your own Spirit severs itself from your flesh-body, then it partakes and uses your own spirit-body. Then you lay aside all mortality and materiality, but you still will be able to operate and function on the plane of the earth in just the same way as you did when you wore a flesh-body. In fact, you will look and appear as though you still had physical body."

(ALLAH is the Supreme Being worshipped by the Mohammedans).

During a materialization session on June 26, 1959, ISH-

MAEL fully materialized himself. He wore a beautiful robe of silver and white and an odd bonnet-type headdress. He said: "I am ISHMAEL. Lara, my child, I have spoken to you through Sara. Can you see me?" I replied: "Yes, Ishmael, I can see you and I am happy that you have come tonight." He said: "I travel to many far distant worlds and I now have come from one of the 'Rings of Saturn' . . . a planet in your solar system. Saturn gives birth to many rings of psychic power, which power now is coming to the earth. I work very closely with the earth, because I am helping to bring peace instead of war. Numerous Beings are now encircling the earth."

August 23, 1962, ISHMAEL spoke through Sara, saying: "I have been on many other worlds since last I visited here. I went to those worlds to analyze the conditions there. I have been given the authority to travel through the heavenly roadways, and to contact those different worlds to help me in my work to make the way straight for this NEW SPACE AGE. Evidently, you have never thought about the methods that must be used in order to bring about this so-called phenomena. That is the reason I am adding this information to your records.

"Not only myself, but there are fifteen-hundred million other Spirit beings selected to go upon all of the millions of other different worlds and planets, to get the people on those worlds ready for this NEW AGE. We are helping by inspiring the minds of people, trying to inform them about WHAT is scheduled in the future for the earth and for all of those other worlds.

"There are millions of advanced Beings who have been living in the different spirit worlds for perhaps millions of years, and now they are being used to open up the vibration and to make the way straight so that this NEW AGE can be accomplished.

"Manifesting this NEW AGE is no ordinary matter. This is something of such an outstanding character that you have never read about it in the history books! You must understand that you are given this information so that you have a record of WHAT IS AHEAD OF THE EARTH WORLD. And also what is ahead of the entire universe and its multi-millions of worlds.

"THE PLAN OF CREATION IS SO VOLUMINOUS AND SO WONDERFUL THAT NO ONE HUMAN BEING CAN IMPRESS ANOTHER HUMAN BEING ABOUT THE COLOSSAL PROCEDURE THAT IS TAKING PLACE.

"This is advanced information, telling what is coming to the people of the earth. I repeat, what affects the earth also affects all of the other worlds."

MESSAGES FROM THE PLEIADES

The constellation of several hundred Stars are called THE PLEIADES. As they are viewed in the sky, from the earth, there are Seven Very Bright Stars that appear to form the outline of Two Dippers. They are known as Ursa Major, the Big Dipper, having five of those Bright Stars, and Ursa Minor, the Little Dipper, having two of those Bright Stars. Those Seven Brightest Stars are also referred to as the Seven Sisters.

Beginning in the year 1957, and continuing for many years, a Highly Evolved Being named THE STAR OF THE EAST, and also known as THE STAR, spoke through Sara. He said that he came from the Pleiades. He brought a great number of messages to us at our center. He also materialized himself before us twenty-three different times during the period of time when materialization sessions were conducted at our center.

THE STAR brought SISTER PLEIADES with him to our center on February 6, 1959. It was her first contact with us. She fully materialized herself that night and also again materialized on May 8, 1959. Her messages on those dates are included in the materialization chapter.

The following quotes are excerpts from a few of their numerous messages that they brought to us when they spoke through Sara.

February 17, 1961, SISTER PLEIADES said: "I come from the Two Dippers that are called THE PLEIADES. They are a mountain-range of many great worlds, where a large number of Evolved Beings live. They have attained a wide range of experiences and wisdom. They are very wise Sages and

Teachers and Masters. I, MYSELF, REPRESENT THE ENTIRE MOUNTAIN-RANGE OF WORLDS CALLED THE PLEIADES. They are using me expressly to tie in with the earth world's vibration, so that they can send their THOUGHTS to the earth people. They are bombarding their THOUGHTS against the thoughts of the earth people, in an effort to raise up the THINKING of all earth mortals. To lift them higher to prepare them so that they can be used for this NEW AGE."

February 20, 1959, SISTER PLEIADES said: "I am SISTER PLEIADES and I use that name for recognition. I come from the Pleiades from the Big Dipper. My Brother, THE STAR, is the Great Personage in one of the worlds in the Pleiades. I have much information to give you regarding the Pleiades."

May 19, 1961, SISTER PLEIADES gave us the following message. It tells WHAT HAPPENED TO THE PLEIADES.

"If you will look at the position in which the Stars are placed that form the Pleiades, how do you suppose they got into that position? I now call this to your attention.

"At one time those Stars rotated and were situated in an entirely different manner. Then something happened to their solar system, eons gone by. Those planets in the Pleiades solar system were on the same level, or were on the same par as your solar system as it is at this time, and as your earth world is at this time. They had people that were just as crude, just as wicked, selfish and greedy, as the people are now on the earth. Then a GREAT CATASTROPHE happened! This affected not only one world, but it affected a large number of worlds.

"WHEN THAT GREAT CATASTROPHE HAPPENED, the location of a number of worlds were changed in the spinning of their orbits. Their locations were changed due to their own manner of living and their actions, and that is why you see them now in the form of Two Dippers. Heretofore, they were stationed in a double circle . . . one large circle and one small circle. During the catastrophe both circles were broken, leaving part of the circle still formed, hanging down, and the other part straightened out. It was done for the reason and the purpose that the whole solar system of the Pleiades did not, would not, and could not destroy itself.

"NOW THIS IS THE HISTORY AND THE STORY OF THE PLEIADES, THAT NO MAN ON THE FACE OF THE EARTH HAS EVER BEEN TOLD ABOUT."

April 22, 1959, THE STAR said: "The TWO SONS OF GOD who are here in this center tonight have come from a world you know nothing about. They have come from a solar system that has a great deal to do with SPACE SHIPS that have been coming to the earth. Many space ships come from the earth's solar system, but there are a great number that are coming from other solar systems and other worlds. Countless far-off worlds are in existence and they are inhabited by many different kinds of Beings.

"Many of those space ships that have been coming to the earth are controlled by Great Beings and they come from those other worlds. They contact this center because it is surrounded by Great Hosts of MASTER TEACHERS who use this center as their meeting place. Those Evolved Beings bring their power here as they are very powerful people."

June 17, 1959, SISTER PLEIADES said: "The earth world has come into the AQUARIAN AGE. This NEW AGE affects the solar system of which your earth is a world.

"WHAT AFFECTS ONE WORLD ALSO AFFECTS ALL WORLDS, BECAUSE THEY FOLLOW A CERTAIN INTERWOVEN PATTERN. ALL CREATION OPERATES LIKE WHEELS-WITHIN-WHEELS."

June 24, 1959 PLEIADES said: "The High Master Teachers who come to this center, if you really knew who they are and the type of Beings they truly are, you would be AWED and have great humility of soul. Perhaps you could not even speak, as it would be so far beyond your understanding."

November 27, 1959, SISTER PLEIADES and THE STAR said: "We are both here and want to tell you our purpose for coming. We have come here with power from our Star Worlds in the Pleiades. We are here to help the Great Ones from our Pleiades Worlds, to bring about their work on the earth. Little do you realize how very important their work is that they now are demonstrating on the earth. They need the channel (Sara) through whom they can come and bring their messages. There is a Great Spiritual Awakening that is coming to the earth people."

April 27, 1962, THE STAR said: "Tape recordings are being made of all sessions when Lara is here, so that you will have the transcripts. In time to come, all of these records will be worthwhile and be very outstanding. Some day earth people will read about events that took place in this little center."

July 3, 1959, SISTER PLEIADES said: "Everything is vibration. In a moment of time, I can leave my Star World and be here in your midst as I am here tonight. We come in an instant, even though we come from a far-off world.

"Many of the High Teachers who have come to the earth from different worlds located in many solar systems, will make themselves known as they come here to this chapter.

"Furthermore, you will have space ships come from those other worlds to the earth. You will find that many of the space people are not Spirit Beings, but that they are mortals, the same as you are. Those mortal Beings are only living in a higher degree of intelligence. That is due to the fact that on their worlds they do not have murders, wars and other such destructive evils.

"I SAY TO YOU: 'DO NOT GIVE UP THE WORK YOU ARE DOING AND REMAIN TRUE TO YOUR TRUST, BECAUSE THE TIME WILL COME WHEN YOU WILL SEE OUTSTANDING PHENOMENA ON THE PLANE OF THE EARTH.' "

MESSAGES FROM THE PLANET URANUS

August 29, 1958, during an afternoon session with Sara, a BEING said: "I am URANUS and I come from the Planet Uranus. This is the first time that I have been present here. I am not the God of Uranus, but I am of a very high office in my world. One of the Golden Age Masters brought me here.

"Now, I will give you a description of my planet. It is a world similar to the earth world. We have made less progress than Venus has made, because Venus is the Planet of Love and on Venus everything manifests in the form of spiritual love. Mortal earth people know a form of love, likened to a love that the animals have for one another, but earthlings know nothing about the rudiments of spiritual love. On Uranus we have gone beyond the animalistic love, but we have not yet come to the place that Venus has developed, whereby Venus people express and function through spiritual love. Every world in existence eventually must operate through spiritual love. Our mentality on Uranus has gone a step higher than earth mankind has gone.

"I now have tied in here with this vibration, with the magnetic belt that has been built here, so that I can help the earth in its journey upward.

"Tonight, I will psychically take both of you and your Master Teachers with me to show you my planet Uranus . . . and how we operate. You will find buildings similar to your own on the earth. However, those buildings on Uranus have a greater consistency where color, sound and music is concerned, because we have come to the place that in our atmosphere we hear a constant soft music playing, and you might liken it to the gentle breezes of the wind on your earth. Our sky is a constant beautiful rosy-hued color."

SETHANTES then said: "I have been on the planet Uranus times without numbers. People on Uranus use the same method of communicating with Immortal Spirit Beings as earth people do, and as we now are doing here at this center. On Uranus they have many more classrooms than you have on the earth. I have gone to their classrooms and have spoken through individuals in the same way as I now speak through Sara."

September 2, 1958, the BEING named URANUS from the planet Uranus, said: "I am here for a purpose, the same as the other Beings who came here from other different worlds in this solar system. There are so many other solar systems in the vast cosmos comprising all of creation, that earth man's mortal mind and my own mind cannot count them.

"I am here to bring you information and also to glean information from the Highly Evolved Master Teachers who come here to this center. I will take their information to people on my planet Uranus.

"On the planet Uranus we do not have wars, which are wrong, sinful and futile. People on Uranus now are tying in with the spiritual love vibration as displayed on the planet Venus. However, we have not yet come fully into that love vibration, which will be the manner and means through which we will operate, work and live, when spiritual love is firmly established on Uranus. Earth mankind cannot come to that place that Venus has attained, until earth people eliminate all wars, all quarreling, murders, suicides and arson, and all of the other evil things now so rampant upon the earth. The earth people have not yet even started on that plane of understanding of spiritual love. That is why communicating centers, such as this one, are established on the earth.

"Communicating centers also are established on the different worlds, so that the Highly Evolved Master and Teachers who come from their different graded, far-off worlds, can come to such centers and bring farther advanced information. The Great Ones usually function through their higher, finer, faster vibrational powers that exist on their worlds. When they come to work on a lower world, such as the earth, they need such centers as this one, to which they can come and dip into and use of the vibrational powers and magnetism that has been built up and stored by the preceding Great Ones at the centers. It helps all of us when we can come to the centers and make use of the stored substances. When we draw from those substances they enable us to more easily live and breathe while we are doing our work on the earth . . . which is invisible to earth people."

July 17, 1959, a Great Being said: "I AM THE GOD OF THE PLANET URANUS. The earth people will be recipients of great demonstrations that are coming to the earth world.

"Russia and China and their satellite countries hold millions of people in slavery and in subjection. All of those countries are being cleansed. And, when the GREAT CATASTROPHE comes to Russia and China and their satellites, you will know that you have been foretold about it in advance.

"This cleansing of the earth is necessary in order to bring into manifestation the New Heaven and the New Earth, which now are functioning through different rates of vibrations. Those two conditions must become harmonized and brought together so that they will function and operate as 'ONE AND THE SAME PLACE.' "

August 7, 1959, a Great Being ISHMAEL, who came from the planet Venus, told us about the noted Great Being named "HERMES." That Hermes who now is working on the earth, came from the planet Uranus. However, Hermes has lived on many different worlds.

Ishmael explained that Hermes and the other Beings who have come to the earth from Uranus bring with them a different personality and a different vibration and power. It is a vibrational power of positiveness, and they do not use any negation. They walk with their heads held high and they have a determination about them which you can feel as it emanates from them. It seems to come through the pores of their bodies and you can almost see the power. That is how powerful they are.

September 2, 1960, several GREAT BEINGS brought their messages to us and the following important excerpts are quoted.

"The earth world is in the throes of a GREAT CHANGE and it is greater than you are aware of, or can comprehend, because events are leading up to a CLIMAX. The vibrations of the whole earth are in a state of confusion. Such outstanding events are coming, both destructive and also constructive! Many things will change on the earth."

"This is the dark before the dawn which is ahead for earth people."

"The Planet Uranus also is going through a period of reconstruction and it is having difficulties, and it is pulling on the earth. The whole solar system is affected."

June 10, 1964, a Great Being made his first contact at our center. He said: "My name is merely for recognition, and I call myself the PRINCE OF URANUS. But I am not from the Planet Uranus in your solar system. However, the Planet Uranus in the ages past, has played a very important part on that higher world that I have been transferred to and now have come from to the earth world.

"I am showing myself to the channel through whom I am speaking. I am the color of red, such as you have never seen on the earth. I am also robed in a robe of red, from the top of my head to the soles of my feet. My garment is not all in one piece.

"This outfit that I am clothed in has many different compartments, but they are not pockets. There are many compartments all over my outfit, covering every inch of it, and each compartment has a purpose. In each one is a gift. They are gifts that the CREATOR is sending to HIS children on the earth, and all people are His children. These are mysterious gifts and you do not know what they consist of. You will be given these gifts, one at a time, and you will utilize them in your mind and in your manner of operation.

"First of all, I will open a compartment and take out the gift I have brought for each of you. The gift comes from the top of my head, upon which there is an object. I want to explain this as it is not too complicated but it takes in a great area. On the top of my head is a nearly square object about the size 4 x 6 inches. Inside of that one object there is something that has never been brought to you before, and I tell you now what it is.

"I have brought from a great world from outer space, what you would call soil, or dirt earth. I now show the soil to the medium. It is the color of rust. Now, cup your hands, as I want the power to come into your hands, Lara, and also into Sara's hands. This soil comes from the great high-raised world from which I have arrived.

"Now, I am placing this soil on top of your head, and very slowly I am penetrating your skull, the bony part of your head, and I am opening a tiny little excavation so that this soil can go down through your skull and travel through your body until it reaches every inch of your body. It will penetrate every atom of your body.

"These gifts that I am bringing to the earth people are what will be used for the CHANGES that are coming to the bodies of all earth people, during this New Age Cycle of Time.

"It is very important that you, Lara, are recording the information I have brought here at this time. Your records will be very instructive to earth people. They contain information explaining what is happening on the earth. Whenever a New Cycle of Time arrives, everything changes on the earth.

"This, that I have brought to you tonight, has nothing to do with UFO'S or SPACE SHIPS from outer space."

JUPITER, SATURN & PLUTO
PLANETS IN OUR SOLAR SYSTEM

June 26, 1959 GOD JUPITER, who is the Chief God ruling the Planet Jupiter, materialized himself before us. His message is included in the materialization chapter.

* * *

June 16, 1959, ISHMAEL, a Great Being, who travels to many different worlds, materialized himself before us during a materialization session. His message is included in the "Messages from Venus" chapter. He spoke about the Planet Saturn, therefore that part of his message is repeated herewith.

"I now have come from the Planet SATURN in your solar system. Saturn gives birth to many rings of psychic power, which power now is coming to the earth."

* * *

August 29, 1958, MATHENO, A GOLDEN AGE MASTER, spoke through Sara, saying: "We, the Golden Age Masters are working on the earth world, and when we finish our work on the earth, we are going to help the PLANET PLUTO, located in your solar system. PLUTO is a very, very backward planet."

(Matheno materialized himself before us on 8-22-58)

MESSAGES FROM THE STAR BEINGS

Several STAR BEINGS from different Star Worlds in our solar system have brought many important messages to us at our center. On May 19, 1961, they gave us the following information.

"We want to assure you and enlighten you upon certain facts. We have come exclusively from the Star Worlds as they are very much tied in with the earth world. We are vitally concerned with everything that takes place upon the earth. What affects the earth also affects the Star World. We have come to help with the rejuvenation of the earth.

Countless Star Beings and other Great Master Teachers have come to the earth from their Star Worlds and other far-off worlds. Those many millions of Mighty Immortal Beings are hovering over the earth. They also are walking upon the earth among earth people. They can appear as earth mortals. However, mortals do not recognize them as Immortal Beings. They can quickly appear as mortals and then in an instant quickly disappear from the physical eyes of earth people."

They have explained to us that all of the Highly Evolved Spirit Beings become magnetized and spiritualized to the powerful faster rates of vibrations that exist in their various graded spirit realms of life in which they dwell in the higher dimensions of the spirit worlds. When they come to the earth they come into the earth's slow, coarse vibrations. Some of those Great Beings find it difficult for them to live and breathe in the earth's vibrations. Therefore, other Great Beings who have preceded them when they too came to the earth, have established many different centers, such as our center. They have brought their Mighty Powers to those centers, leaving those powers at the centers so they can be used as "dynamos of power." Those Great Beings who need to harmonize their faster vibrations with the earth's slow vibrations seek out the centers where they can dip into the established powers to help them live and breathe so they can perform their assigned work on the earth.

All of the Great Beings are bringing to the earth their higher, faster electrified vibrations. They have helped to build the magnetic belt that has been placed around the earth for the earth's protection. The earth also will operate through that magnetic belt during this New Age. They now are working hard to keep the earth on its axis.

The STAR BEINGS told us that the GREAT GOLD STAR BEING who has come to the earth will be one of the Rulers over the earth when this NEW GOLDEN AGE is firmly established.

This is a list of the names of the most Prominent Star Beings who brought many messages to us.

BRIGHT MORNING STAR, STAR OF THE EAST, EVENING STAR, NORTH STAR, BETHLEHEM STAR and THE STAR who came from the Pleiades solar system.

The GOD OF THE UNIVERSE who brought a great number of messages to us, spoke through Sara on April 10, 1963. He said: "It is needful that you know what is taking place on the earth so that you will have the information in your records.

"All of the High Raised Master Teachers, including the many Star Beings, who have come to the earth and made themselves known to the people at this center are here for one purpose. All of them are working under "ONE PLAN" which comes directly from the CREATOR.

"This GREAT EVENT which is scheduled to take place on the earth, is also reaching into all other worlds. It is a revitalizing and revolutionary demonstration to bring all worlds together in harmonious spiritual operation. It is the PLAN OF THE CREATOR."

June 15, 1961 the EVENING STAR Being said: "All of the STAR BEINGS play a very important part in helping to bring the CREATOR'S PLAN into manifestation."

WORLD OF ESMERALDA

LIKA was the Chief God who ruled the earth during this last cycle of time. It was called the Cycle of Bon and it ended about the year 1849 A.D. Then this present Aquarian Kosmon Age Cycle began.

On April 9, 1957, GOD LIKA spoke through Sara and said: "I am psychically taking both of you with me in a space ship to the WORLD OF ESMERALDA. It is located in the high etherean spirit firmament that exists above the earth's atmospherean firmament. The World of Esmeralda is a spirit world.

"The space ship that we have entered is circular and has no windows, but it is very spacious and is white inside and we are sitting on a circular upholstered divan. Now, we are going through space at a terrific rate of speed, like a bullet shot from a gun. Quickly, our ship has landed on the world of Esmeralda and we step out of the ship.

"We come to a beautiful garden and see the lovely green grass and the trees, and far in the distance we see an ocean. 'IT IS THE OCEAN OF LIFE.' Let me explain this to you. An Ocean of Life exists on every etherean world that the CREATOR has created. The people living on those etherean worlds do not experience so-called death, which is a transitional phase of life which earth people experience.

"As you look into the etherean atmosphere you see many different colors and also see a blue sky, all dotted with what appear to be silver and gold stars, but they are not stars; they are other worlds.

"Now, I will take you to one of the great auditoriums which seats many millions of Individual Spirit Beings. You will see the manifestation of the CREATOR'S creations in the faces of those myriads of Souls. You see a beauty there that is reflected in those faces which mortal man's mind is unable to grasp. The bodies of those Soul-Beings are composed of all Light, shining forth all the different colors.

"I now take you to show you the flowers which are of such heavenly beauty, sending forth their delightful fragrance.

"Now you see two very large beautiful white birds, with crimson breasts and crimson tails and they have very long legs. They are as large as an ostrich that lives on your earth. I am going to surprise you by having these two birds speak to you, because they have a language and they can speak even as earth people speak. They may not be able to give you very much conversation because I have to build the vibration so that they can make their thoughts manifest in speech. I now place my hands on their larnyx and say to them: 'Speak to these two earth people and greet them as best you know how.' The two birds in unison said: 'We are two creations who come to you from the CREATOR, Who has given us life and Who has given you life.' Lika said: 'Thank you for the demonstration.'

"Now I take you into a beautiful banquet hall and its walls are transparent but they look solid. Those numerous Souls that you have come amongst, they are so far spiritually advanced that they do not see you, because they are so glorified and so glorious. I wanted you to have a glimpse of the GLORIFICATION OF THE CREATOR.

"We now go back to the earth in our space ship."

URILLIUS, A Great Master Teacher at our center said: "I and many other Great Masters went with you on that space ship trip. We helped to hold the vibration in order to accomplish the space trip."

(Lika materialized himself before us on 3-6-59 and Urillius materialized on 6-13-58 and on 11-27-59)

The Teachers have explained to us that there are countless other solar systems and worlds that lie far beyond the etherean worlds.

RUBEON PLANET

On September 19, 1958, a Great Being named RUBEON, from the Planet RUBEON, made his first contact at our center, saying: "Earth children, this is RUBEON and you know nothing about me. I come for a very important purpose, to bring power to this medium, Sara, which she needs.

"When we, the Teachers, come to mortals at the centers we bring power to strengthen them and bring concrete information for their enlightenment. We do not waste valuable earth time to bring trivialities. We do not operate through time on our planet.

"I say to you, all people must eventually operate through the POWER OF LOVE, FOR LOVE IS THE CREATOR."

September 26, 1958, RUBEON said: "I come to the medium but I am a Universal Worker and I have tied in with the vibrations at this center for a definite reason. I come from the Planet Rubeon. I have brought with me an Associate named PHILO, who is a Golden Age Master. He lives on the Planet Rubeon when he is not working on other worlds and planes."

Then PHILO said: "The Universal Teachers and their great numbers of Co-Worker Hosts, draw from the vibrations of all earth people who attend the various centers. But your own special personal Teachers draw vibrational power from their students exclusively. When a Teacher comes to talk to you he stays in your vibration as long as he can draw from you the power he needs for him to breathe while in your vibration.

"All earth people are living in a vacuum and are functioning in a substance called magnetism, which you do not understand, as you do not realize what it consists of."

OTHER FAR-OFF WORLDS AND THEIR PEOPLE

During our many sessions, the different Great Beings who came from their far-off worlds, gave us information about their worlds and other galaxies of solar systems.

On September 12, 1958, a MASTER TEACHER said: "The Etherean, Pyramid, Triangle, Triune, Crystal Worlds and Love Worlds are solar systems of their own. Each such solar system contains many thousands of worlds that are located in separate groups in one of the roadways in the Universal Cosmos. There are so many solar systems with their millions of worlds, that the mortal minds of earth people cannot grasp their countless numbers and their descriptions, or the descriptions of their inhabitants, because they are so fabulous."

May 17, 1957, PO, a Master Teacher at our center said: "Today, Sara and Lara, I am taking you on a psychic trip. We are going in a space ship to a world which is only one of the many millions of worlds operating in the roadways in the high etherean spiritual firmament.

"I am taking you in 'Thought' and in 'Spirit' to a GOLDEN PYRAMID WORLD. I want you to see and to feel the different atmospheres we will pass through. Understand that 'Spirit' can pass through any atmosphere, and it may be 100 degrees below zero, and the 'Spirit' would not be affected. Only earth people are affected by changes in the atmosphere. This is a trip through the ethers of space. I will point out many things to you and you will follow me in your 'Thoughts.'

"First, I am taking you through the stratosphere, through a different hemisphere, what you would call ice-land. As you look down from the ship, many fathoms deep, you see ice caps and they come up in points. They are glistening white as the sun shines on them and they look like great myriads of jewels having many colors. And as you look farther and deeper way down deep into the bowels of the stratosphere, you will see a glorious blue sky and sunlight down there. You will see the green fields, trees, mountains and oceans. Now, we are going down to that world and stop there for a little while. Before we could reach that hemisphere we had to go through the ice-glades formation.

"As we land our ship, a great host of etherean people are awaiting our arrival. As you look out before you step down from the ship, you will see a sea of many faces and as you look into those faces they are not like the faces of earth people. Each face looks like a very delicate flower. There are no features like eyes, nose, mouth or ears. These are different faces and they are very beautiful faces. They look like beautiful flowers, such as your pansy flowers portray faces. The texture of their faces and their bodies are not like flesh. They are entirely different. As you look at that great array of faces they look like a garden of flowers, though they are Individual Beings and they are the CREATOR'S Creations.

"Those people do not come near enough to touch you because you are operating on a different rate of vibration, due to the flesh body to which you are attached.

"Now, we have come back on the ship again and as we go a little higher, as you look down over the terrain of that world, you will not see buildings like those on the earth. Their buildings are just a visible color that you can see through, and those people live in those buildings like those on the earth. Their buildings are just a visible color that you can see through, and those people live in those buildings when they so desire. The buildings are not divided into rooms. The outside of their buildings is not like ground or soil like you have on the earth. Their sky is a panorama of many colors. Their different colors play a very important part in the world, as everything is done through colors. The people show their affection through the colors vibrating around them. When they contact a Being whom they love very much they emanate their loving colors.

"They do not need to eat like mortals do on the earth. They merely breathe-in the colors. The colors are in their atmosphere and in their vibrations. Those people subsist and live on, or through, those colors. They are a very happy people. As those people are waving 'adieu' to us, they are not waving their arms, but they are waving colors that go out like streamers to all of us who are on this space ship. We are not alone as there are many other Master Teachers who are our companions.

"Now again, we are fastening our ship on a plateau just above the WORLD OF ETHEREA. It is an ethereal world that you can see through every obstacle. You can see through space for millions of miles. You can see the landscape, the beautiful flowers, trees and an ocean, somewhat like those things on the earth, but there really is no comparison. You also see little transparent animals that are very playful, joyful and lovable, and they can speak if they care to speak. They speak audibly in sounds, as they do not speak with thought. The GODDESS OF ETHEREA greets us and takes us into her home.

"Now, I have taken you to two different Etherean Worlds, but I have left the greatest demonstration until the last.

"Again, we are speeding through space at a very rapid rate of speed that is faster than the speed of THOUGHT.

"Now, we have come to this PYRAMID OF GOLD that has a golden light surrounding it. It is the symbol of the world I am taking you to, a GOLDEN PYRAMID WORLD. We are anchoring our space ship a few miles above it. That is as close as we can come to it today, as the vibrations here are too high for Sara at this time. So we will depart and return to the earth, and bring Sara back to normal again."

(The many groups of PYRAMID WORLDS operate through their own solar systems. They are located far beyond the millions of ETHEREAN WORLDS).

LOVE WORLDS

Our Master Teachers at our center have explained to us that a plateau of many LOVE WORLDS belong to the same solar system that contains all of the millions of PYRAMID WORLDS. The Love Worlds lie far beyond the millions of Etherean Worlds.

The LOVE WORLDS are the LOVE PLANES of the CREATOR'S Creations. The GREAT INFINITE LOVE THAT COMES FROM THE CREATOR is of such heat intensity that earth people in their physical bodies would not be able to stand the vibrations of the Love Worlds.

May 25, 1957, a Great Being, named the BRIGHT MORNING STAR, who was one of our Prominent Master Teachers, spoke during my session with Sara. He said: "I am taking both of you on a psychic trip to the Love Worlds. We are not going in a space ship. We are taking you there in your spirit-body and are traveling swiftly by the POWER OF THOUGHT.

"Now, as we arrive at one of the Love Worlds and as we stand at the edge of that world, we cannot see beyond or around it, because it is a structure of great magnitude. And as we stand there, I want you to feel the PEACE and the QUIETUDE, that all outside things in your mentality are closed out of your mind. All that you see inside as we go into that world, is its ETHEREAL BEAUTY. All things are ethereal, which you cannot touch, but you can only feel them as they are spiritual. You can feel, see and hear things in that world. You can hear the faintest, most beautiful music. That music is brought about by the colors playing back and forth through that world.

"As far as your eyes can see you will see the beauty of living forms. They do not walk but float through the air. They simply move through their atmosphere by the power called SPIRITUAL LOVE. It is LOVE alone that they send out. That is what causes them to float about in the air. They touch one another as they have that GREAT SPIRITUAL LOVE for the one they touch. Earth people must realize that the attribute of Love really is the SPIRITUAL LOVE OF THE CREATOR. In the Love Worlds, there is where the CREATOR'S LOVE demonstrates to its fullest capacity. That GREAT SPIRITUAL LOVE that those Love Beings emanate, they send that GREAT LOVE out into the atmosphere. Those Beings are ethereal Beings and they are the CREATOR'S Creations on those SPIRIT LOVE WORLDS."

June 28, 1957, PO said: "Recently, I took both of you, Sara and Lara, on a psychic trip (5-17-57) to two beautiful Etherean worlds. At that time you also saw a glimpse of a magnificent GOLDEN PYRAMID WORLD. On May 25, 1957, the BRIGHT MORNING STAR Being took both of you to one of the GOLDEN PYRAMID LOVE WORLDS. It belonged to the solar system containing all of the many thousands of Pyramid Worlds. I want to enlarge on that information.

The Pyramid worlds are composed of several groups of formations of worlds shaped like a Pyramid. That LOVE WORLD the BRIGHT MORNING STAR took you to, is in the shape of a Pyramid. It is only one of many thousands of Pyramid Worlds.

The Highly Evolved Spirit Healer named TES, who does great healing work from this center, came from a GOLDEN PYRAMID WORLD. I now ask the different Great Beings who are here to speak to you."

The BRIGHT MORNING STAR Being said: "Today, we are tying in with another PYRAMID LOVE WORLD. It is white, the color of quartz crystal. It is thousands of miles long and thousands of miles wide, and hundreds of miles high. It consists of vast space. IT IS INHABITED BOTH INSIDE AND OUTSIDE WITH GREAT BEINGS.

"We stand at the entrance of that world. It glistens like a diamond. We can see what is there in that world and we can look through it. We can see for millions of miles as though we were looking through a jewel of many colors.

"In that world are trees, flowers, grass, oceans, mountains, valleys and buildings. But they are not stationary, because they move and have their existence in the ROOTS OF THE CREATOR. All things in that world are constantly moving. The trees move to another location, at will. The ocean, in the twinkling of an eye, changes its position and location and then functions in a different part of that world.

"That crystal clear white diamond LOVE WORLD is inhabited with people having very high luminosity and very high vibrations. Each person is one solid color. All of their different colors are so outstanding that it is impossible to describe them. They are color and color alone. All that you could see is perhaps a beautiful red, or blue, or some other color and that person functions in that color. There is a body but not like your earth body. It is a body of such high vibrations that in your imagination it would almost seem like nothing at all except color. Or, if you would see a sunbeam coming through a crack of a door, you could pass your hand through that sunbeam. That is as near a description I can give to you.

"The people too, are not stationary, because their vibrations are so high that they constantly move. Now, you earth people in your phase of life, would not be able to understand that, as you would think that would be very unsatisfactory to live that way. Each person on that world can go from place to place in the twinkling of an eye. They can go where they want to go, finding themselves in that very place. How do you suppose that rapid traveling is done and manipulated? It is done alone through the POWER OF THOUGHTS. Each person THINKS themselves into a certain part of that world, and the instant he visualizes himself at the place he wants to go to, he is there. He takes his building (home) along with him in his THOUGHTS. That magnificent crystal clear white diamond PYRAMID LOVE WORLD is very much larger than the earth world.

"No two worlds function and operate in the same manner and through the same vibrations."

(The BRIGHT MORNING STAR Being materialized himself before us on 2-28-58. He said: "I show myself to all in this room. I have a star on my forehead." His beautiful headpiece also had a shining star on it. He was tall in stature and was very elegant).

PYRAMID WORLDS

December 27, 1957, the MOON GOD said: "I am taking both of you, Sara and Lara, on a trip to the PYRAMID WORLDS where I lived before I was appointed to rule the Earth's Moon. I am taking you in your spirit body.

"Now, we are on a plateau which has a brilliant golden sky that constantly shines like your Sun when it rises in the morning. The magnetism surrounding that sky is of such great power! As far as we can see this golden sky is prevalent and there is no night there and the blue stars are constantly in the constellation. Those blue stars eventually will become worlds.

"I now touch your third-eye in the center of your forehead. It is your spirit-eye. As far as your spirit-eye can see you will see many millions of Pyramid Worlds. Those worlds have a color all of their own and they have all the colors of the rainbow. No two worlds have the same color of luminosity. Those Pyramid worlds appear to be flaming because their colors are sending out electrical sparks of luminosity, due to their high degree and quality displaying spirituality.

"We now have arrived on a plateau from which we view a group of Rose-colored worlds having the shades of Pink, Rose and Red.

"As we now move very rapidly beyond the Rose Worlds, we see the group of Pyramid worlds having the colors of Amber, Yellow and Gold. That group has so many worlds that they cannot be counted. As we look afar out, we can see the group of worlds that have the shades of Light Blue, Royal Blue and very Deep Blue worlds that are vibrating with those colors. We look farther on and see another group of worlds consisting of many different shades of beautiful Green colors. I also point out to you the group of worlds having the different shades of Lavender and Purple colors.

As we stand at a point where we can look afar out, we can see all of these different groups of beautiful Pyramid Worlds with their scintillating, brilliantly shining colors that are throwing off electrical sparks of luminosity. The Great Beings living on those worlds are Electrical Beings who are greatly charged with electricity. Earth people have only a small amount of electricity in their flesh bodies.

"ALL LIFE IS ELECTRICITY!"

(The GOD RULER OF ALL THE PYRAMID WORLDS materialized three times).

A FAR-OFF SPIRIT WORLD

November 14, 1958, during my session with Sara, METHUSALAH said: "I have journeyed here from afar, and you may think I could come here in the twinkling of an eye, which is absolutely true, as I can do so. But it is far better for me to glean experiences and greater knowledge as I loiter along the way. I can tie in with the different vibrations on various worlds. I appear before many Great Beings who govern those worlds and I garner information that I can use in the work for which I have been assigned.

"I have not lived in another physical body since I left the earth world, as an Individual known as Methuselah. I lived in earth life more than 900 years and ENOCH was my earth-father.

"Right now, I am taking both of you, in your spirit body on a trip to a very beautiful home in a far-off spirit world. The atmosphere there is a golden color. Everything in the home is the color of gold, glistening brightly. There are many different musical instruments, and now comes someone who has a beautiful flow through his fingers. He sits down before a lovely harp and his fingers glide over the harp strings as we hear heavenly music.

"We, who dwell in the spirit realms, can hear many Great Orators and Great Musicians. We hear a great display of harmony. The beauty that we behold is on such a colossal scale that we cannot begin to describe it.

"We can become bathed in the CREATOR'S GREAT WHITE LIGHT, which attaches us and every Human Being to the CREATOR. Then as we see ourselves in that GREAT LIGHT, we can drift off into the Great

Beauty of Understanding.

"We can go to the Spirit Love Worlds and to the Pyramid Worlds, because we tie in with many of those worlds. We can go there as rapidly as THOUGHT. It is the same as though you turned a page in a book and you saw a beautiful scene, then as you turn that page, you see an entirely different scene. Those experiences are what we now are going through and you are with me in your spirit body. We have briefly tied in with the LIFE OF SPIRITUALITY. You have seen a glimpse of the Great Beauty existing in many Spirit Worlds. I have fulfilled the mission for which I was sent to you today."

MIZPAH ORGANIZATION

On May 23, 1960, during my session with Sara, a Great Being said: "You may refer to me as MIZPAH as that is my name used entirely for recognition. I belong to an organization called MIZPAH.

"There are many millions of different organizations in the realms of spirit life. They are formed by groups of Spirit People who band themselves together for the purpose that they can bring power and magnetism to earth people.

"I am going to describe myself. I have a body but it is a different body from what you can imagine a body. It is a body that can become enlarged and take in a great area of space. Or, I can diminish myself into nothingness and disappear. I wear a cloak of many colors and my garment is composed of the same nature and texture as my body, for I can add or diminish the colors of my raiment and the volume of it.

My face, if I can call it as such, is not a face like you have. It is entirely different and I will not go into any explanation. On my head I do not have hair such as you would know it, nor would I say that it is a headdress. But I say it is a white substance with one wave after the other from the beginning of my forehead until it goes down below my shoulders. I also have a full beard. I have in my hand a gold wand. I pass this wand over your heads to give both of you greater strength."

SEVEN GOLDEN AGE MASTERS

Beginning in June, 1951, the Highly Evolved SEVEN GOLDEN AGE MASTERS made their first contact with Sara. Then during the many years that followed, they continued their contacts with her. They told her that they have lived upon the earth in physical bodies during many past cycles of time. That they now are on the earth in their spirit bodies, as they have been assigned to perform a very special work upon the earth. Shortly after my sessions started with Sara (Yr. 1957), they were Prominent Master Teachers at our center. They brought many NEW GOLDEN AGE messages to us. Their names are:

APOLLO, who, many past ages ago, lived on the continent of Lemuria before it was submerged in the Pacific Ocean. He ruled the earth as its Chief God, 18,200 years ago. He now is on the earth in his spirit body, working as the "Grecian."

MATHENO, the Egyptian; PHILO, the Hebrew; MEGESTE, the Chinese; KASPAR, the Persian; ASHBINA, the Assyrian.

FRAGAPATTI at one time lived on the earth in the land now called India. He ruled the earth as its Chief God, 8,900 years ago.

The Golden Age Masters explained to us that each one of them were assigned to a major nation on the earth. They were to try in every way possible to work with the Leaders of those nations. To inspire them with altruistic thoughts and try to teach them the righteous ways of governing their nations.

They further explained to us that they have come to the earth to help bring into harmonious operation and manifestation this present NEW GOLDEN AGE upon the earth. The heavens of the earth and the surface of the earth must become blended together as "ONE AND THE SAME PLACE" so they can harmoniously function through higher spiritual vibrations.

Several of the Golden Age Masters materialized themselves before us. We directly talked with them and also walked with them. Their messages to us at that time are quoted in the materialization chapter.

GREAT BEINGS BROUGHT MESSAGES

The Highly Evolved GREAT BEINGS have explained to us that our entire Universal Cosmos and all that it contains is a harmonious electrical system of countless mighty powers. All Human Beings are electrical Beings. Each person operates through his own magnetic vibrations.

January 17, 1958, a GREAT BEING said: "All solar systems and their worlds operate through their own magnetism and vibrational electrical powers. All Spirit Worlds have their separate graded plateau levels, or planes, or dimensions, where their Immortal Spirit Beings live.

"When we travel forth from our far-off worlds and desire to visit other higher worlds, we must become magnetized to the electrical vibrations operating on those higher worlds. We can only stay on them for a short time if we cannot harmonize our vibrations with those existing on those worlds. We may find that it is difficult for us to live and breathe while we are there and we may become dizzy and our vision may become blurred. We cannot remain there if we have not attained the 'Breath of Life' that is operating on those higher worlds.

"Many of those other higher worlds and their higher plateau levels, or dimensions, we cannot reach. First, we must develop ourselves and progress through the proper conditions. Those conditions are made through a certain spiritual growth and a proper chemistry must take place in our body. Most of all, we must have the wisdom in our MIND. We can attain that knowledge from the much Higher Evolved GREAT BEINGS, who lovingly teach all people who desire to rise higher in their wisdom and greater understanding of

LIFE, as it is lived on the countless higher worlds.

"That is exactly what I now am doing. My own MIND is opening up much wider and I listen and learn greater lessons that explain far advanced GREAT SECRETS OF LIFE, from the GREAT HIGH TEACHERS who come to this center from their higher worlds."

November 20, 1959, a group of GREAT BEINGS, who referred to themselves as the "GREAT ONES" made their first visit to our center. When the GREAT ONE NO. 1, the Leader of the group, started to speak through Sara, she received a sudden "shock" of electricity that was very powerful and consisted of higher magnetic vibrations. She very loudly exclaimed: "OH-H-H-H." Then the GREAT ONE said that they had come from a world located in the SEVENTH PLATEAU OF SOLAR SYSTEMS. He also said, that still other Higher Solar Systems lie far beyond their SEVENTH PLATEAU.

Then he said: "We have been sent to the earth for a special purpose. This is the opening wedge, the beginning of a GREAT FABULOUS DEMONSTRATION that is coming to all people on the earth. With 'One sweep of my POWERFUL MIND' I shall help to bring a decided change in the way people are now living on the earth. I am here in my 'three' different bodies."

After the session ended, Sara said: "When that GREAT ONE took hold of me and started to speak through me, his great power took me over and my chair vibrated. I have never felt such strong vibrations. I thought we had an earthquake."

November 27, 1959, the GREAT ONE NO. 1, fully materialized himself. He appeared in an earth-type human body form and he wore a beautiful garment. In his right hand he held UP HIGH A TORCH WITH A FLAMING LIGHT IN IT. He said: "May your paths by paths of Peace. Always hold high in your hand the TORCH OF TRUTH AND LIGHT. Can you see the torch with its flaming light?" We replied: "Yes, we see it and it is beautiful. Thank you for coming to us and demonstrating the torch."

December 18, 1959, the GREAT ONE NO. 1, said: "This

TRUTH that is brought to you, and the work that the GREAT BEINGS are accomplishing on the earth WILL NEVER FAIL."

Then Sara described a GREAT ONE who psychically showed himself to her as a GREAT POWER BEING. He did not give his name, or speak.

Sara said: "I see a GREAT BIG BEING in this room and I will try to describe him. He does not have a form as we know human forms. He is moving about and I can only describe him as a GREAT BIG POWERFUL BEING but I cannot describe his features or his body. He looks just like he is a big cloud and there is an opening in that cloud and I see the outline of a Being but his features are not like our features. This demonstration is so far beyond what we now can undertand."

January 8, 1960, the GREAT ONE NO. 1, said: "We have been building up Sara's vibrations, whereby we can have the HIGHER ONES speak through her. Also, to enable us to spiritually take her to visit the higher worlds.

"Now, I am taking her in her own Spirit to a magnificent world in the SEVENTH PLATEAU OF SOLAR SYSTEMS. She sees the matchless grandeur of the buildings that are of great magnitude. Their heights are indescribable. Earth people have never seen such buildings on the earth. Now, we go to the top of the tallest building and she feels very lightheaded. I will try to describe that building as briefly as I can bring the information through her mentality.

"It is not composed of materials such as you use on the earth. All of those buildings are made of beautiful gems and jewels, some that you have never heard about. They are composed of all kinds of precious gems, such as jade, diamonds, rubies, emeralds and other gems. Those gem-stones are many feet high and many feet wide and you can look through the gems. They are as clear as crystal because they do not have a blemish. They glisten in the colors of which they are composed, reflecting such colors as ruby-red and other beautiful colors. The different colors throw out their brilliant color-beams into the universe. We also hear the wonderful harmonious music that comes from the atmosphere. It is glorious music such as earth people have never heard.

"This description given to you describes only the outside of the buildings. It is all that I now am able to show to this channel at this time. Perhaps at another time we will take her to the inside of those buildings. I also at this time shall not show you the GREAT BEINGS that are in those buildings. I only wanted to describe the GLORY and the GREAT MAGNIFICENCE of those SEVENTH PLATEAU WORLDS and the half has not yet been told. I have described only a faint glimpse of the beauty that is in those worlds. This is only one such world that I have described to you.

"This is only the opening wedge to explain to earth people the great beautiful things that lie ahead of them as they journey through life during this New Age Cycle of Time."

After the session ended, Sara said: "I am still feeling dizzy and my head is swimming. That was the most outstanding thing I have ever seen since I have been psychically taken to other worlds by the Great Master Teachers. This is the epitome of seeing what those Seventh Plateau Worlds consist of."

SPHERICAL . . . A MIGHTY MASTER

On January 7, 1959, a Great Being named SPHERICAL made his first contact at our center. He arrived from the Triangle Worlds Solar System. He psychically showed himself to Sara and she described him. "He is extremely tall and he wears a very tall headdress that appears as though it is heavy and it touches the ceiling in this room. Around his head is a crown of Stars and they do not touch each other. His face is quite long and he looks different from what we mortals look like. He has very large, dark piercing eyes and a black beard. The garments that he wears are covered with all sorts of things, like jewels and other things like gold and silver. I have never seen anyone laden down with such things as he has on his garments. He is enveloped in a beautiful dark blue haze. He has never been here before this time. He has brought with him five small-sized people who are his Helpers. They are only about four feet tall and they are dressed in dark clothes. One of them holds something like a compass in his hand. It has a needle and the needle moves."

Then SPHERICAL said: "I refer to myself as an 'Image' rather than as a Master. I have come from a solar system you have never heard about. My name that I will use here at this center is SPHERICAL . . . S-P-H-E-R-I-C-A-L. (He spelled it).

"I am now going over the earth and will appear in those centers that have been established by other Masters who have preceded me. There are millions of Triangle Worlds in the galaxy that I have come from to the earth. I do not recognize time and do not know how long it took us to come here.

"The earth has entered a NEW DAWN and a NEW DAY.

"We speak a different language than earth people speak." (Then he spoke in a language that sounded like this).

"BIA—BA—BIA . . . BIA—BIO—BIO—O—O—O—O." He said: "It means, the CREATOR Bless You and keep you in all HIS Ways."

"The five small helpers, or Images . . . I do not call them Spirits or Masters, they came in a conveyance, and I do not call it a space ship, because it is something that conveyed all six of us here into your midst.

"I say to those people who can visualize me and my Helpers, *that* is imagination and imagination is Spirit. I hope that people on the earth will begin to realize what spirit language is.

"I have come to give much information to the people of the earth. This is my first stopping-off place, and from here we will go to other similar centers. I have come with a new vibration and have tied in with the vibrations of this instrument (Sara).

"We have come to the earth bringing 'Something' through the atmosphere. We are planting 'Seeds.' I call them 'Hormones' but they are not the hormones that the earth people speak of. We are seeding the atmosphere, preparing for the demonstrations of psychic phenomena, of what earth people will see and experience in the future during this New Age Cycle of Time.

"These five little Helpers that I have brought here are not very tall and their bodies are small, but they have an 'Intellect' and 'Intelligence' that few people have. Their intellect, stamina, courage, determination and love that they have for all of the ALMIGHTY CREATOR'S Creations far exceeds their small size.

"The world in the Spherical Triangle Solar System that we have come from is far higher than any you have tied in with. The people on the earth are now being made acquainted of WHAT is coming to the earth. Many STRANGE THINGS will take place. I want you to know WHAT is taking place and how those things are brought about. I am here to tell you distinctly of what we are building here, so that at no time are you left in a mental state of confusion, doubt or fear. If those doubts would creep into your mind then they would be detrimental in our work. We must work through harmony and love.

"As the days come and go, you will tie in with more Highly Evolved Masters than myself, who will come to the earth from farther worlds in other solar systems."

January 9, 1959, SPHERICAL said: "Reach out for the higher knowledge that is in your MIND and in your THOUGHTS, because the information that is brought to you is brought through the MIND and the MIND IS THE PART THAT IS CONNECTED WITH THE CREATOR. Week after week, there will be much added astounding information brought to all of you people who work here at this center with us. I shall come here many times and bring information during the next following years." (He then spoke in a language that I could not transcribe)

Again, SPHERICAL spoke: "This is the beginning of things that are coming to the earth, as the NEW HEAVEN and the NEW EARTH are becoming established. When the NEW EARTH fully manifests, that is the time when the mortals on the earth and the Immortals in the spirit world which functions with the earth, will mingle together and dwell together as one happy family."

Then TES, a Great Master Healer, said: "The Great and Shining One, SPHERICAL, as I see that 'Halo of Stars' about his head, I count the stars, for I know they have a special meaning, indicating his attainment of very high spiritual evolvement. Only when a world, such as the earth, is ready for a spiritual upliftment . . . to be brought into higher spirituality . . . only at such a time, does this GREAT ONE, SPHERICAL, go to such a world. He has been doing this for many thousands of years. He works with the CREATOR and now he is here on the earth. That is proof that the earth has passed her emancipation and in the future will operate through the spirituality after it is thoroughly cleansed of all wrong things now so rampant upon the earth."

January 16, 1959, SETHANTES said: "This GREAT

BEING, SPHERICAL, is more highly evolved and farther advanced that I am. He has attained greater wisdom. In the days to come other Great Masters will also come here who will be even more highly evolved than SPHERICAL.

"A band of magnetism enfolds the full body of a Great Master, and SPHERICAL is building a band of magnetism around each person at this center. Tonight, he is sending out across the room, his rays of magnetic power covering your bodies. He and his millions of Spirit Hosts will also do this for other mortals."

On January 16, 1959, SPHERICAL said: "I am tonight working through Sara's entire spirit body. I am using the mentality of her spirit body, because I am unable to work through her flesh body. It is too gross and too immature for my work. I was unable to manipulate the odic force in her flesh body. Now I am using the odic force which is generated and powered through her spirit body. Now she is quite comfortable, so that I can breathe more freely and more easily doing my work.

"I am analyzing and testing Sara's vibrations. When I have harmonized her vibrations so that they will function freely with my own higher vibrations, then I can bring to you psychic visions and information, the like of which you have never seen or heard.

"This is not just ordinary manifestation of myself, as I work through Sara's body. She must be so magnetized and so electrified that you might be able to see the electric sparks as they come through her fingers.

"These are the things that must take place in order for me to demonstrate the work for which I have been sent to the earth.

"An individual person can be described as a vibration. The vibration from another person can envelop the vibration of someone else, and that vibration can influence an individual.

"The conditions that must be made to enable us to bring about demonstrations, must be made through the magnetism and the electrification that comes from your own flesh bodies. The flesh body must be magnetized by us and become like a great dynamo of power for us to use. From that body will go sparks of electricity, though unseen by mortals. It is that electrical substance that is necessary to be used in order for us to demonstrate many things that will be brought into manifestation for physical eyes to behold.

"Mortal man has very little magnetism and electricity in his flesh body, because it is not made up of the atoms that are conducive to electricity. The flesh body is a very gross body, compared to the other bodies that compose and are used by the Immortal Beings who live on far-off higher worlds in the spirit realms of life. If earth mortals would have a sufficient amount of magnetism and electricity, then many material objects near them would become weightless and they would float in the air.

"A Great Master can build around himself a magnetic belt of power so that mortal people cannot come within the radius of that power. Because of this 'belt of magnetism' that is built around him, people are not able to approach him. It is an electrical magnetic belt that repulses them. If people would touch him they would take away some of that power infolding him. He has that power around him for his own protection, and with that power surrounding him he can make himself invisible.

"Now, I will bring something else to your attention. During the time when the ISLANDS OF ATLANTIS existed on the earth, the people on Atlantis operated through a magnetic belt, such as I now am making possible here. Many things were demonstrated on Atlantis through the use of magnetism and a magnetic belt, such as their mode of living and their mode of travel. They used those methods and brought them into existence, and they lived a happier life during many centuries. The people on Atlantis operated on a Principle and now we are bringing that same Principle here to the people on the earth. We are starting here in this vibration.

"I have been sent here to the earth to magnetize the flesh bodies of mortals, so they will be able to put heavy things into their places without the aid of human power of mortal Beings, doing such things simply by using the magnetic force that surrounds them.

"These are some of the NEW AGE DEMONSTRATIONS OF FUTURE EVENTS that are coming to the earth. They are so fantastic and so far-reaching that mortals cannot now understand them. It is the MIND of the individual person and the way that MIND operates that determines what message can be brought through that individual's vibration. It is the individual's MIND entirely through which we work. The way a person's mind works and gives out, that is exactly what we have with which to do our work.

"When an individual's mind wanders, then it cannot be held steady. His mind must be held to a focal point and become steady until it does not waver. His concentrated attention must be focused upon the speaker's message. If possible, that person should at all times try to picture in his MIND'S EYE what is being relayed to him. His mind must be held steady so that it does not take on other thoughts and other conditions. This is a natural law which I have brought to you today."

(Spherical materialized himself before us on 3-6-59).

GOLDEN CHARIOT BEINGS
(Riders In The Sky)

During my session with Sara on July 29, 1960, the Great Being named ROCK OF AGES, showed Sara a psychic vision which she described: "I see the sky is filled with white-robed BEINGS who are riding on white horses. The riders are all 'muffled-up' and they wear white headdresses with a long cloth that hangs down to their shoulders, like the Arabians wear. They are riding very rapidly and there is a great wind blowing. They tell me they are from the PLANET OF THE STARS.

"Following those Riders there are chariots that are white trimmed with gold and they have white horses. The BEINGS driving those horses that are pulling the chariots, wear voluminous garments with decorations that look like the ancient 'Walls of Troy' insignias. As I look into the faces of those drivers of the chariots, their skin is red like the North American Indians, and their hair is very black. In their one free hand they each have a long pointed spear, like a javelin, that contains mighty rays of power, and they are throwing those javelins. They tell me they have surrounded the RADAR MISSILE BASE at Thule, Greenland, as they are there to guard that part of the earth's hemisphere."

(The Ballistic Missile Early Warning System, with its mighty radars, was built by the United States, England and Canada, to guard the free world from sneak attacks).

On November 7, 1961, Sara again was shown a psychic vision which she described: "I see many GOLDEN CHARIOTS in the sky and every chariot has a white-robed MAS-

TER BEING in it. The chariot vehicle is part of their body and it is a conveyance that propels them."

Then, a GOLDEN CHARIOT BEING spoke: "We operate in the vehicle that was shown to the channel. Each chariot has wings and those wings are movable, and we operate the vehicle by our THOUGHTS. That is how we propel ourselves through space. We live in the atmopshere and live in a different manner than the way you live. We do not require food and other things of life such as you earth people require.

"Our mission at this time is very far-reaching. We are here to help cleanse the earth of all its wrongful acts. They can only be erased by destructions, such as catastrophies. Not one thing happens by chance or is done in vain. HAVE NO FEAR!"

SEVEN GREAT FEMALE BEINGS

February 9, 1962, a Mighty Being named UNKNOWN ONE, said: "In time to come, there will be SEVEN FEMALES who will come to this center. They will tell you their purpose for coming to the earth. Each female has a very important assignment."

August 3, 1962, the Great Being named the HOLY ONE, said: "The LADY OF DESTINY, who is the first (No. 1) Female, has now arrived at this center. She is a very beautiful, Highly Evolved Immortal Female and this is her first visit to the Earth.

"In time to come, she will play a very important part in government affairs of all nations. She will be seen in the assembly room of the UNITED NATIONS and in many other prominent government offices. She will be responsible for many drastic changes that will take place in the High Offices of the Government of the United States. The masculine part of the Human Beings, who have been ruling the governments, have failed in their ruling ways. Now, the Feminine Beings will come to the fore and take over the ruling of the nations. She will place a woman at the Head of the United Nations."

LADY OF DESTINY said: "I have been especially assigned to the United States of America. I have come to the earth to bring about many necessary changes in your high government offices. More women will become elected to high offices in your government.

"There will be six other Great Females who will come to the earth and take their parts in world affairs. The females will help to bring into manifestation this NEW GLORIOUS

GOLDEN SPIRITUAL AGE. Immortality and mortality are now coming together as ONE PLACE . . . AS ONE CONDITION. This is the opening of an entire new life for earth people.

"The CREATOR is bringing about MANY GREAT WONDERFUL THINGS AND ASTONISHING DEMONSTRATIONS that are coming to the earth 'from out of the air.' This revelation is far beyond anything you could possibly imagine at this time. There is a GREAT DAY coming and this NEW DAY . . . this NEW AGE is upon you. Prepare yourselves for the GREATEST COLOSSAL EVENT OF UNBELIEVABLE MAGNITUDE THAT IS COMING TO THE EARTH."

April 22, 1964, Sara described the SEVEN HIGHLY EVOLVED FEMALES whom she psychically saw in the session room.

"NUMBER 1, who is the LADY OF DESTINY, is a very beautiful person and she wears an elegant robe. She has fine chiseled features and appears similar to us earthly females. The other females do not look like she does as they are different types of Beings. Now, the Lady of Destiny changes the colors that are around her and she changes her appearance, and also changes the beauty of her robe. Now, she has stretched out her arms and is extending them upward. There are sparks, like fire-crackers, that sparkle and glisten, shooting out from her fingers. Then, with her right hand she covers your head. Perhaps Lara, you may feel the spirit heat because she is very powerful."

LADY OF DESTINY said: "We females have come with our millions of Great Associate Hosts, who will help us accomplish our work on the earth. We have been Rulers on High Raised far-off worlds from which we have come. We have practiced the teachings of the CREATOR and have become imbued with Great Wisdom. We have been prepared, during many thousands of years, for this period of time. We will play a very important part where the earth world is concerned.

"This is why we Seven Females have come to the earth, to give greater impetus to the women on the earth. We are bringing our power here and it will reach out into all nations on the earth. In the future, women in every country will take part in their government politics, showing forth their capabilities in the enactment of new righteous laws. They will hold high offices in all of the governments on the earth."

NUMBER 2 FEMALE. Sara said: "She too is very beautiful. She is the LADY OF HARMONY and is also named BLUE STAR or BLUE RAY."

LADY OF HARMONY said: "All things must harmonize and operate in perfection. Mortal people are composed of several different bodies. I have come with my numerous Spirit Hosts to harmonize the different bodies of earth people so that they will operate in a perfect harmonious condition during this NEW AGE."

NUMBER 3. Sara said: "She is sitting on a Golden Throne and she looks as real to me as though she is here in physical embodiment. She is a peculiar looking person. I do not clearly see her face, but I see her moving her arms and they are somewhat like ours. But instead of having five fingers on each hand, there are two projections of each finger. They start from the palm of her hand as one finger and then the finger splits into two fingers, so that each hand has ten fingers. She has no face such as we have and it looks like a Radiant Sun-Ball. There is a body, but what would be the head and the face is a ball-of-light shining like a Sun."

The SUN LADY said: "I am the female whose countenance shines like a Sun. I do not appear with features such as you mortals have. My body is something like your own but my face is like a Golden Sun.

"The Golden Throne is symbolic of the position the United States will hold in the NEW AGE. I shall be at the helm of the Females who will rule the nations of the earth, when the time arrives for the females to rule. I will be stationed in Washington, D.C., and be responsible for the righteous laws that will be passed.

"I have attained great wisdom and I have the Voice of the CREATOR, that will be heard throughout the earth when the proper time arrives during this NEW AGE."

NUMBER 4. Sara said: "As I look closely at her, she has

TWO BODIES that extend down to the torso. There are two legs and two feet. From the waist up there are the TWO BODIES. I see TWO HEADS that look alike and they are covered with what looks like wool. I see no features. She tells me she is a PSYCHIC BEING, bringing psychic phenomena to the people on the earth."

PSYCHIC LADY said: "I have a great task to perform—*To open the psychic powers of each person on the earth.* Their psychic powers have been dormant during their entire earth life. Now, my millions of Associate Spirit Hosts are opening that "odic psychic power" to bring it into manifestation. Many people will realize that something strange is happening to them."

NUMBER 5. Sara said: "She is sitting down and appears in the color of yellow. She does not look like Human Beings as we know them. She looks like something I can't describe as her whole outline is peculiar. Now I see her body which is very large and looks like it is formed from what we would call a vegetative substance. It looks like what we call little bushes, or little trees and they form her outline. She looks very queer."

PROTECTING LADY said: "I have come with my millions of Spirit Hosts to protect many thousands of people on the earth when great catastrophies happen."

NUMBER 6. Sara said: "She is very beautiful and as "black as ebony." She has brought with her a black Male Associate. They are helping to bring about progress with the black people on the earth."

BLACK BEAUTY LADY said: "My Male Associate and myself represent the black race of people in Africa who just now are beginning to get their freedom from slavery. They too, are children of the CREATOR. Now, they will be given the opportunity to become more civilized and educated. We, together with our millions of Associate Spirit Hosts, will place knowledge and wisdom in their minds and rejuvenate their minds and their bodies. The black people must be raised up to the same level as the white people on the earth."

NUMBER 7. Sara Said: "She is a very Magnificent Beauty. She wears a beautiful headdress which is as large as this session room. It is covered with symbols which represent many different things. I see no face. She appears as ONE BEING, yet she appears as NUMEROUS BEINGS. I can't describe how I see her, but I know she is ONE PERSON, and yet I see thousands of BEINGS. However, she is still only ONE BEING. I know that description of her does not make sense, but that is the way I see her. She is beyond description."

STRANGE LADY said: "During this NEW AGE you will begin to see many strange things and strange happenings that are coming to the earth. MANY GREAT SECRETS OF LIFE will be revealed to earth people, that now are unknown to them. My very strange way of manifesting myself to you pertains ONE GREAT SECRET OF LIFE. The future fabulous demonstrations are now beyond your comprehension."

(On September 25, 1981 Sandra Day O'Connor was the first woman who was appointed as a Justice of the Supreme Court of the United States of America.)

PURPLE RAY BEINGS

On March 11, 1964, the PURPLE RAY BEINGS made their first contact with us at our center. Sara said: "I see several Great Big Beings and they are very tall and their garments are made of solid looking different kinds of jewels. I cannot count how many of those large Beings are here in this vicinity and some of them are walking around in this room. I have never seen these Huge Beings before tonight. They are coming from some place that we have never tied-in with. As they surround me, they are so powerful that they almost have taken my breath away from me."

Then MICHAEL, who is One of Sara's Master Teachers, said: "We will try to diminish the power as much as possible. These Great Beings who have come from their higher worlds, have brought their greater power with them. The mighty electrical powers coming from these Highly Evolved Beings affect the physical body of this channel. We will see to it that the power of these Great Ones will not overwhelm her. As time rolls on, Higher and More Highly Evolved Beings will come to this center. Each new group that arrives bring their greater power, and they also use more power from Sara. They throw their Great Power over her when they use her as their transmitting channel. Tonight, as these Great Beings covered her with their Great Power, it flooded her with unusual sensations in her head and in her mind.

"These Great Ones indicate by their magnificent garments, the high advancement they have made on their pathway of life, having dwelled upon many different highly advanced worlds and planes of existence.

"As you have been told, many groups of Beings from far-off worlds are coming to help the earth, because the earth itself now is being raised into higher vibrations of power. It now is operating in the Higher Spiritual Arc of Kosmon. These Great Ones have come to bring their greater wisdom to the earth people.

"A PURPLE RAY BEING is standing beside Sara. She now is reaching out her hand and is touching a jewel in his garment. It is an amethyst, a clear purple color. As she presses the jewel with the third finger of her right hand it unlocks and opens up and shows to her a world she has never seen before this time. Now, we quickly take her to that world which is located far-out in space, near where the CREATOR, Himself, dwells. Briefly, she sees a glimpse of that world, and we return her to the session room.

"The Higher Worlds in existence are located near the CREATOR, as HE draws power from them to carry out HIS Plans for the lower worlds, such as the earth. That is the reason why these Great Purple Ray Beings have come here tonight."

Then Sara described the PURPLE RAY BEING whom she closely saw standing beside her. "As I look into his face, which is utterly indescribable, I can only describe one feature at a time. He has a HUGE HEAD that is much larger and much longer than any head that you could imagine a Human Being would have. His HAIR (what you would term hair on a head) is NOT HAIR. It is a substance that has no substantiality, for if you would touch it you could not feel it with your physical fingers. That substance on his head looks like flax and it has a light straw color."

Then the Great Purple Ray Being began to speak through Sara. He said: "Good evening! I wish to describe myself. I think I can give a clear mental picture to these earth mortals.

"My FOREHEAD is used for many different purposes.

"MY FOREHEAD APPEARS AS A CONTROL-BOARD-CENTER.

"I HAVE FOUR EYES and FOUR EARS.

"Every so-called hair on my head is an antenna.

"If you could touch my FOREHEAD, you would touch a CONTROL-BUTTON, and it would tie-you-in, and it would open your mentality into many, many different worlds, upon which life is functioning in a way such as you have never heard of. This is only a minute glance of Life as it is lived on some of the highly evolved worlds.

"There are many buttons, gadgets and levers in my forehead. You can either pull them out and look far beyond and see great worlds. Or, you can press them in, and then you would see a form of life which you have never anticipated, even in the wildest dreams of your imagination. My forehead is much wider and much deeper than you can possibly anticipate.

"MY EYES: I have what you could call EYES, but there again, they are not eyes as you have eyes in your physical body. I might say they are a lense that can reach afar out and cover a large territory. My lenses take in many things of such a nature that you could not now comprehend if I told you about them.

"MY NOSE: I do not have a nose such as you have. I do not need a nose through which to smell, for I have no use for the function of that organ.

"MY MOUTH: I do not have a mouth. You may say: 'How then, are you speaking through a mortal Being?' I say to you, I am not speaking through a mortal Being. The mortal herself is speaking, but I am providing the material . . . the subject matter, through my THOUGHTS. I am not using the larynx, the throat, or the tongue of this channel.

"MY EARS: I have TWO EARS ON EACH SIDE OF MY HEAD. They are very large and they help me to LISTEN-IN and hear what is taking place on many distant worlds. At this very instant while I am now on your earth world, I can listen-in to many far distant worlds. When I am on my own world from which I have come, then I can listen-in to every world in existence. Every world, regardless of where that world rotates in its own orbit, I can listen-in and tune into the MINDS and THOUGHTS of every HUMAN BEING living upon those worlds. I CAN LISTEN AND HEAR AND KNOW WHAT THEY THINK AND WHAT THEY SAY. I can control an Individual Person who is living upon one of those worlds. I can control them by the power of my

THOUGHTS. I can impress upon their MIND what they are to do and what 'Truths of Life' they are to impart to Other Beings, giving them knowledge pertaining to OUR HOLY CREATOR, according to HIS instructions to me, pertaining to HIS GREAT COSMIC PLAN OF CREATION THAT IS EVER UNFOLDING.

"Now, from my description of myself, you must readily understand that I have come from a very High Plane of Life into your presence. I know you have closely followed my description of myself, as I have given it to you, Lara. I watched the beams of light emitted by you as they played across your consciousness and how they played through your aura. I watched the lights as they flashed on-and-off, as you pictured in your mind exactly what I was describing. Now you have comprehended and have a mental picture of what I look like.

"For recognition only, I will call myself the PURPLE RAY, or the VIOLET RAY. Now, I have placed that Purple Ray into the atmosphere and it goes out expanding itself throughout the atmosphere.

"I now am drawing this Purple Ray Power from my body to the solar plexus of this earth channel, for a definite reason. Her flesh body is very important where we are concerned, because we must use a mortal Being who is dwelling in a flesh body as an instrument through whom we must do our work. We came to the earth to perform many things for the benefit of all mankind. I have come to the earth to do my certain work that has been assigned to me to perform."

Then he connected a "Beam of Light" between himself and Sara. It was a Purple Violet Ray that tied them together.

He said: "She is living in this earth life and I am living in my state of life. Between her and myself, I have connected this Purple Ray, which ties us together. I am doing this because of the work that is ahead of the whole earth world. Sara is 'One Individual' whom we can use in order to bring our work into focus and into operation, and we must do it in this manner.

"Our work links-in with all of the other things that are necessary for the rejuvenation of the earth, getting things ready for all the people who will live upon the earth during this New Golden Age. All of the necessary things must be prepared in advance, according to the COSMIC BLUE-PRINT, when the rebuilding of the earth starts.

"Tonight, we have laid out the Blue-Print Plan, and now everything is placed in order, getting ready for the rebuilding of the earth and the 'Building of this New Age.' That work now will go forth on a greater scale.

"IT IS THE NEW BUILDING OF THE EARTH AND OF THE NEW HEAVEN SURROUNDING THE EARTH. THE COMBINING OF THE TWO IS OF SUCH GIGANTIC PROPORTION, THAT IT WILL REQUIRE A LONG TIME, TO BRING THEM INTO HARMONIOUS FUNCTION IN THEIR ENTIRETY.

"I am now placing in Sara's body and in your body (Lara), this Purple Violet Ray that not only connects Sara and myself, but it now also connects both of you mortals together. This now, is the 'dynamo' that we have prepared for the new work that Sara will enter into. It is necessary, first of all, to 'season' the instrument (Sara), before the information can be brought through to the other people who belong to this small group at this center. Now tonight, we have made the connection. Also, this entire room is now sensitized with this Purple Ray.

"As I have just told you; 'I can pierce the innermost "Thoughts" of every Human Being living in every world that is in existence.' Naturally, I know everything that has taken place at this center.

"I am still standing beside this channel (Sara). Now again, her hand is touching another jewel in my robe, which is made of all the different kinds of solid jewels.

"The planet upon which I live is known as the PURPLE RAY PLANET. Many worlds in existence are known as 'RAYS' and their different rays are sent forth wherever the CREATOR instructs those rays to go, in order to enchance and advance life on certain planets. THAT IS THE PRINCIPLE OBJECT OF THOSE RAYS, TO ENHANCE 'LIFE ITSELF' AS IT IS LIVED ON ALL THE WORLDS IN EXISTENCE.

"I have come for the express purpose tonight to acquaint you with my phase of the work that will go forth from this center.

"Each Human Being must redeem himself from every short-coming he may have developed while in physical embodiment. When the group of Mighty Beings, called the REDEEMERS, came to this center (1-4-63), they came here to teach people on earth how to redeem themselves. To help them to bring themselves back to their original state of purity. The REDEEMERS function through the GREAT LOVE POWER, WHICH COMES DIRECTLY FROM THE CREATOR, and it infolds all Human Beings in existence.

"I have come with my Great Host of Evolved Beings, who will work to help to increase the Positive Power of Love and expand it throughout the earths. We will work to abolish hatred among all mankind on the earth. Our work includes removing all the evil, negative vibrations on the earth. You need both the positive and the negative vibrations on the earth, but the negative vibrations have out-balanced the positive vibrations, and these must again be counter-balanced. Earth people usually link-in with the positive love vibrations and work through those vibrations, or they tie-in with the negative hate vibrations and work through them. And they walk in the so-called darkness of spiritual understanding of Life.

"We are preparing the earth for the NEW AGE, by eliminating all evil, negative vibrations. In order to accomplish this work, it requires many changes that must take place, working through MIGHTY BEINGS such as myself, and through mortals such as yourselves. The combined forces are the ones that must bring these changes into manifestation. This is the work that now is taking place, and tonight the FOUNDATION OF HOLINESS has been established, and now our work will go forth through a much higher demonstration.

"A GREAT VISTA IS OPENING UP FOR THE EARTH WORLD AND A GREAT CHANGE IS NOW TAKING PLACE. And as you WATCH and WAIT to see these changes manifest, you will remember that the Purple Ray

Being advised you of those changes, and that my Great Host of Purple Ray Beings are working through the Purple Ray. We have set the pattern for these things to be brought into fruition.

"This is my message to you tonight, but I shall come again. And, as I leave you, I will not leave you empty-handed. I now pluck a jewel from the right side of my garment, and Lara, I lay this spiritual jewel in your hand. It is an amber, yellowish colored jewel, with high brilliance. This is a very high spiritual color and it is neither yellow, nor is it a golden color, but it has many yellowish colors combined within it. For you, Sara, I give to you an onyx jewel. It has shades of many colors which are symbolic."

Then, as the Purple Ray Being began to depart, Sara uttered a loud shriek, saying: "OH . . . OH . . . OOOOOO." Then she explained to me what had happened. She said: "Gee, that was like a bolt of electricity when he left me. It was an unexpected shock to me. I just now have opened my eyes which have been closed. I have been mentally awake during this long discourse by the Purple Ray Being."

Then, A Highly Advanced Being, named GOD OF THE UNIVERSE spoke through Sara: "This Great Purple Ray Being will now predominate at this center. He will gradually explain the work in more detail. He has come with his Great Wisdom and Knowledge. It is difficult for me to explain what he really consists of, with all of his own built-in gadgets that he displayed in his forehead. They are like radios and televisions with antennas. They represent all the electronic inventions pertaining to the communication services, both mental and electrical, that are displayed throughout the universe, which link-in with all of the other worlds in existence. THESE ARE THE MOST MARVELOUS AND OUTSTANDING INVENTIONS that have ever been brought to the people of earth, and he is establishing their connections upon the earth.

In rapid succession the many mortal inventors will invent new types of gadgets and machines that will typify the things that the Purple Ray Being has displayed tonight in his forehead. When these are brought into manifestation for constructive purposes, you will remember how they have been brought about.

I want to bring this to your attention very forcefully. The work that the Purple Ray Being performed here tonight will change the entire earth world. It will change every nation on the earth and the way each nation operates because it will change the principles under which they operate at this time. You will discover that all of these changes, from time to time, in rapid succession, will take place and manifest as we have predicted.

"Both of you will be used for many different purposes but you will not be conscious that you are being used. The Great Beings silently draw the necessary substance and power from each of you. They do this all unnoticed by yourselves. You will know why many unusual happenings and inventions of astonishing new things manifest because you have been told in advance about them.

"In the past, many Great Holy Ones have come here to this center, telling you of their missions and of their work they are performing on the earth, helping to prepare it for this New Age. Tonight, greater impetus has been given to this 'gigantic work' and henceforth the work will proceed much more rapidly.

"All of the many predictions telling about future happenings to take place on the earth, that have been explained to you, they now are beginning to take place. You will marvel at what you will see and hear.

"The MIGHTY PURPLE RAY BEING brought with him millions of Associate Co-Workers who are on the same evolved level as himself. He is their Leader and spokesman.

I have great compassion and humbleness for all earth people, more than I have ever had before, because this Great Purple Ray Being has brought a new vibration with him, which will enter into the Minds and Thoughts and bodies of all Human Beings living on the earth.

"We, the Evolved Great Ones, now are working to cleanse the earth of all earth-bound Spirit Beings who are living in their former earth homes. Such Spirit Beings are called 'ghosts.' We also are removing from the earth those millions of Spirit Beings who lurk in the dens of iniquity on the earth.

"LIFE IS ETERNAL AND LIFE NEVER STANDS STILL!

"All Human Beings must be taught the rudiments of 'LIFE' as 'LIFE' really exists.

"There are unlimited 'MYSTERIES' pertaining to Life, which are ever ahead of all mankind, awaiting to be solved.

"When mankind realizes how many millions of GREAT BEINGS comprise the 'HIERARCHICAL GODHEAD' that governs ALL CREATION, then they will have an extended outlook upon LIFE. All mankind will stand in AWE and MARVEL at the existence of our OMNIPOTENT ALMIGHTY CREATOR . . . THE DEITY, WHO ALONE KNOWS ALL THE MYSTERIES OF LIFE."

When the session ended, Sara said: "All the time when the Great Purple Ray Being controlled me, I was somewhat in a daze. This really was an outstanding session tonight." I replied: "Yes, it was astounding and almost overwhelming. I always am amazed how the Highly Evolved Beings speak through you and control their rapid, smooth flow of words during such a long session. They clearly explain to us why they have come to the earth. To help to bring PEACE, LOVE and HARMONY to all earth people."

SEVEN RAY BEINGS

On Novembr 26, 1964, Sara said: "This room tonight has the color of yellow. Lara, your new YELLOW RAY TEACHER is demonstrating the different colors of yellow in this room."

Then, the YELLOW RAY BEING spoke: "The time now has arrived when the RAY BEINGS will demonstrate their 'rays of power' through their various shades of colors. That is why the RAY BEINGS have come to the earth.

"There are SEVEN OF US RAY BEINGS who work together as a group. WE ARE SEVEN RAYS and we are not Beings like you mortals. We do not have forms or figures but we are RAYS OF LIGHT. We only perform in rays of light and we move in circles and in cycles so that we can help to bring about the 'new activity' that is coming to the earth.

"Visualize me as the YELLOW RAY. The other RAY BEINGS are the Blue, Red, Green, Rose, Lavender and Purple. Each Ray Being reflects his different shades of colors. We perform in our various shades of light. Many demonstrations on the earth will be performed through the Light-Ray Colors."

GOD OF THE UNIVERSE said: "As we merge closer and closer in this New Age, many drastic changes throughout the earth world are taking place. And, as we draw nearer to the year 2,000, that is the time when this New Age will begin to operate through many new conditions."

Another Great Being said: "I am the SAGE OF THE AGES. The ages can be of antiquity, or be of the future ages ahead—for all mankind. New trends of THOUGHT are now being brought to earth people. And, as many mortals receive these new strange trends of thought, and new ideas, they become confused. Each day the psychic powers of many mortals increase, causing them to seek and search for greater enlightenment pertaining to the spiritual phases of life.

"The people on the earth are being rejuvenated, as well as the rejuvenation of the earth itself. Whatever affects the earth, also affects every planet in your solar system. The ALMIGHTY CREATOR Himself is constantly expanding and growing, therefore, all Beings are likewise affected and must also grow.

"Life everywhere is changing and new ways of doing things are brought about. In due time, there will be no death of the physic body, but all of mortal man's other several bodies must be changed, and raised into higher vibrations to conform with the functioning of this New Age.

"Many people coming from other planets and worlds will visit the earth and you will meet them on the streets. Some earth people will be caught-up in space ships and taken to other planets. Many space people are now operating on the earth but most earth people do not recognize them as space aliens. Do not fear them as they have come to help mankind.

"Earth people are living in a very exciting time . . . watching the changing scenes throughout the earth world. Remember, the CREATOR works through Law and Order, and when the proper time arrives, then all things are brought about according to HIS Plan."

Next, AQUARIUS spoke: "I AM AQUARIUS . . . THE SPIRIT OF THE AQUARIAN AGE. I work through this channel (Sara) and I envelop her mind and her body, because 'she is One of the Great Ones' who are operating on the earth at this time. No matter how hard or how painful it may seem to her, this work must go on. When her bodily changes that now are taking place, when they are completed, she will have entirely new bodies through which she will operate. I have a different body from yours, but I have a face and a form.

"I shall go forth throughout your entire solar system and go out over the whole universe, bringing my vibrations. I was assigned to my position as AQUARIUS millions of years ago and I have been working towards this NEW AQUARIAN AGE during all time.

"We, the Great Ones, are working with might and main, getting the earth ready for its emancipation from its low, slow, coarse vibrations. It now is spinning in its orbit as it travels through the Spiritual Arc of Kosmon. In the future, the earth will become a bright, beautiful work and operate among other spiritual worlds. We are working to help bring about the necessary vast earth changes, according to the CREATOR'S Plan."

Then SETHANTES said: "The entire earth must be cleansed of the millions of evil Spirits who roam about on the earth. This must be done before this New Age can fully manifest. Many Highly Evolved Beings who are working on the earth, are lending their power, helping to bring about these vast earth changes that now are taking place. I say to both of you, be faithful and true to your assignment on the earth, as the Great Ones need you as 'dynamos' through which to work."

MORFIA BEINGS BRING MESSAGES

On February 15, 1963, a Great Being named MORFIA—I made his first contact at our center. He said: "The earth world has entered into a NEW SPIRITUAL AGE, and at this time all earth people must begin to develop their own inherent spiritual faculties to prepare themselves for the exciting, astonishing spiritual phenomena that in the future will be demonstrated all over the earth. Each earth person must reach that future higher spiritual plane of living. All people must reach that higher place in their own mind and in their own thoughts. This will come about very rapidly because many millions of Highly Evolved Beings are coming to the earth to raise up all mortals.

"The majority of mankind on the earth do not know about the drastic changes that are taking place. They only realize that destruction is rife upon the earth. I say to you, there will be destruction of one form or another and many people will lose their earth life and make their transition into spirit life in great numbers."

February 16, 1963, MORFIA—I made his second visit to our center. Sara described him: "I see a GREAT, HUGE BEING, who mentally tells me he is MORFIA—I (the first). There are three such Beings in the room, Morfia—I, Morfia—II and Morfia—III.

"Morfia—I is standing in the middle of the room and he is so tall, much taller than the room ceiling and the ceiling is extending upward. I have never before this time seen a Being as tall and as large as this Being. As I look more closely, I see his face but it is not as we know faces.

"HE HAS ONE BIG TELESCOPIC EYE IN HIS FORE-

HEAD AND ONE IN THE BACK OF HIS HEAD. I do not see any mouth, nose or ears.

"I see these Three Beings who have a forehead and they have hair that seems to be golden. THEIR FACE IS A GOLDEN LIGHT. They have just ONE EYE IN THEIR FOREHEAD, and it extends several feet afar out. They mentally tell me the reason for that extended eye is that they are so far advanced that they need this extended eye, because it helps them to see much farther into the many different worlds."

MORFIA—I then spoke: "I want to take the hand of each one of you, and as I feel the substance of a flesh and blood body, you do not feel my hand because it is transparent. MY WHOLE BODY IS NOTHING BUT LIGHT ALL THE WAY UP TO WHAT YOU CALL A FOREHEAD. There will come a time in the far-off distant future, when each of you will look exactly as we do. Then at that time you will be coming to a material world such as the earth. But the earth then will have become a highly evolved world.

"I will explain the purpose of our visitation. We have come again tonight, simply to test the vibration and to test your mind and your thoughts.

"I say to you, in the future your earth world will not be the same world. Many things will have become changed and a great number of things will be destroyed. Many different new things will be built. Numerous unheard of things will be in operation in the atmosphere and they will operate in connection with other worlds because the earth is going to operate side by side with some of the other higher worlds. The earth now is moving very rapidly into a much higher dimension in the universe. When you move into a higher dimension then the things that have been operating in the lower dimension must be dissolved and destroyed."

MORFIA—I GIVES A VERY GRAVE PREDICTION!

"I now am talking to you of destruction that consists of destroying the wrong things that are taking place on your earth. Also, as you know, there are many countries in your world and several of them will be completely destroyed. You will find that some of the small countries that are scattered all over the earth will sink beneath the waves of the oceans. Other small countries will be lifted up and destroyed in the atmosphere.

"You have read that during the past ages of time, there existed on the earth the ISLES OF ATLANTIS and the CONTINENT OF LEMURIA and they were destroyed. I say to you, those countries did exist and they were destroyed. Also, the destructions of ancient civilizations took place.

"That is exactly what will happen to some of the different islands and countries now existing on the earth. They are going to be destroyed, and the people living upon them will be removed to other worlds. Many people will lose their earth life but they will not lose their spiritual lives. They will merely discard their earthly flesh body and then continue living in their spirit body in a spirit world.

"We came here tonight to bring this information to you. Do not become alarmed, fearful and afraid because there will scarcely be a day pass but that some group of people will be taken off the earth to the spirit dimensions surrounding the earth. That is a natural procedure that the CREATOR has planned and HIS Plans are Perfection. HE does the things that are needful for all worlds that are brought into manifestation.

"The terrain of your world is going to be changed. Where now there are hills they will be leveled. Where now there is water there will be dry land. And that which now is land will be water.

"Time to us does not exist so I will not give you a definite time when these world changes will happen. But this is what is in the process and it is beginning to sweep the whole earth. It now is starting at the bottom of the oceans and it will come to the top. I cannot give you in detail exactly how it will be brought about, because I am speaking through a mortal person and using her as a channel of transmission. I am using her mind and her brain and I am bringing to you the information that I can express through her. I now say to you: THERE IS A GREAT CATASTROPHE COMING TO

THE EARTH AND ALL MANKIND SHALL SEE THE HAND OF THE CREATOR!

"All of these people who will be left upon the earth, and I refer to those who have not believed in the existence of an ALMIGHTY CREATOR, and have no spirituality in their minds, who depend entirely upon their own personal self to accomplish things, they will run and hide behind rocks and in the caves for protection as they will be filled with fear. When they see the lightning and hear the loud thunder they will think the earth is coming to an end.

"This is what is ahead of your planet earth, and that is why we have been sent here and have brought you this information. I will come again and bring you more information.

"This is our first venture upon the earth and when we leave here we will go all over the earth. We will analyze what is taking place upon the earth, and how the drastic changes shall be brought about. We, the THREE MORFIA'S will bring about the assignment given to us. WE WILL START THE CATASTROPHE! At that time you will know that we gave you this advanced information."

PART 4

MICHEL NOSTRADAMUS' MESSAGE

PRESIDENTS OF THE UNITED STATES BROUGHT MESSAGES

MICHEL NOSTRADAMUS' MESSAGES

NOSTRADAMUS was born December 22, 1503, in Lyons, France. He died July 2, 1566. He was a most extraordinary prophet, clairvoyant and astrologer. A printer in Lyons, France published his numerous quatrains of prophecies. He attained worldwide fame when his prophecies came true as he had predicted, averaging nearly one-hundred percent accuracy in his predictions. The scope of his prophecies extended far into the future.

On November 3, 1960, NOSTRADAMUS made his first contact at our center. He said: "This is Nostradamus. As you know, I have been and still am a prophesying medium. I could prophesy things that would take place hundreds of years ahead. But I do not need to predict now what will take place one-hundred years from now, because your Great Teachers have already given you those predictions. They have come from different worlds and they are much higher evolved than myself.

"I have been here at this center many times without you knowing anything about it. I have been listening to the discourses, and I have met Great Beings here that I have never expected to meet before I came here.

"As I have progressed in the world of spirit, I believed that the seventh dimension of the spirit planes that surround the earth was the highest dimension, or level that I could reach in the spirit realms. Then I came to this center and I learned differently, and I have been coming here ever since my first visit here. I say to you that the Great Ones who have come here are beyond compare. Never, in all of my esoteric work, have I heard the things that I have heard right here in this center.

"I know that the things that will be brought here to you in the future are far beyond anything you have had so far. I also will be here, for I know when something 'outstanding' is going to be brought here, because I know by the stream of light that I see coming from this center. Then I avail myself of the possibility of being present. I then receive that same information directly from the 'Higher-Ups' at the same time you are getting it. The Master Teachers send that information out to all of their different centers.

"I want to thank both ADONTUS and PO, who are the Masters in charge of this center, for permitting me to speak here."

NOSTRADAMUS, on August 3, 1962, said: "I will bring prophecies here from time to time. I will not give them in quatrains, but speak to you in your own language.

I say to you, next year in 1963, there will be many changes in your government that nobody knows anything about at this time. There is something that is going to take place in Washington, D.C., and it will cause a great deal of comment. It has to do with the president and his brother. THE PEOPLE OF THE WORLD AT LARGE WILL NOT UNDERSTAND WHAT HAS TAKEN PLACE.

"I shall spend much of my time in Washington, D.C., and I shall come here and relay information to you."

(His prediction did happen. John F. Kennedy, President of the United States, was shot November 22, 1963, at Dallas, Texas. He died that same day. His brother, Robert Kennedy who was the Attorney General, five years later, also was shot on June 5, 1968, at Los Angeles, California).

On Aug. 23, 1962, NOSTRADAMUS said: "This is the New Aquarian Spiritual Age, and many new things are now coming to the earth, and there will be many changes taking place.

"You have been told there will be a NEW MONETARY SYSTEM, but it will not come about suddenly, it will be a long drawn out affair. It is a principle that will operate in every nation. The money will be of the same denomination in all of the nations on the earth. But the money will not have the power it always has had and is having now, because money has been the cause of much of the wickedness on the earth. People will do evil things for money. They will rob and murder people for their money, because money has that power over them. *That* money power will be dispensed with."

NOSTRADAMUS, on August 24, 1962, said: "From my prophetic viewpoint, I bring to the earth people information they should know.

"YOU ARE GOING TO HAVE A GREAT DISASTER THAT IS COMING TO THE EARTH!

"When this great catastrophe happens, the people of the earth will be awed and wonder if the end of the earth has come. I say, it is the changeover through which earth people have been operating. Something new will take its place.

"AFTER THIS GREAT CATASTROPHE, then mortal people will be willing to accept this changeover of WHAT is coming to the earth.

"If this disaster did not come, the peoples of earth would continue in their own way as they have always done. But because their ways have to be changed, is the reason this catastrophe must take place.

"The MINDS of earth people must be changed and they must have a different outlook about life. They must know and understand that new things are coming to the earth, and that the old ways of doing things are finished. A NEW DAY is going to take over and operate. This is why the vibrations of love and understanding must be brought to a high level to earth people. That is the seed that must be sown, so that it can bear the fruit of Love, Peace and Harmony. Then you will have the beginning of the NEW HEAVEN and the NEW EARTH."

On September 8, 1962, NOSTRADAMUS said: "You will have very, very cold weather all over the earth world, including Southern California."

PRESIDENTS OF THE UNITED STATES

Several past presidents of the United States who now dwell in the spirit realms of life, brought messages to our center. Some of their messages are very lengthy, so it is necessary to quote excerpts of their speakings.

ABRAHAM LINCOLN, President 1861-1865. (He was shot April 14, 1865).

On August 29, 1958, LINCOLN said: "I have a message I want to bring to the people of earth. Your history books tell the history of my life and the conditions I had to face. Had I not had the consolation of the Immortal Spirit Teachers from the world of spirit, I never would have been able to carry on in face of all the criticism I received. After I came over into the spirit world, I realized the ways and the wherefors of the criticisms.

"Now I am working with mortals on the earth to help to reduce and obliterate the wrongs that are going on in the earth world.

"We in the spirit world do not use material objects to do our work. We use spiritual power entirely. It has no substance and it is not solidified. It is sunbeams, light, music and color. We get the spirit power from the "thoughts" that people 'think.' People reflect their high spiritual thoughts in their own aura of colored light which surrounds them. And that reflected spirit power helps us to do our work."

* * *

THOMAS JEFFERSON, President 1801-1809.

On September 12, 1958, JEFFERSON said: "I have never been here before this time and I was brought here, and I have been given the privilege to speak. I say to you, the CREA-

TOR controls all things in His vast creations. HE allows mankind to go just so far doing unrighteous things of life, and then he calls a halt. However, HE uses righteous Human Beings through which to manipulate HIS Power to do the work HE wishes to be performed."

* * *

FRANKLIN D. ROOSEVELT, President 1933-1945.

On September 19, 1958, ROOSEVELT spoke: "When I was in a physical body, I did the best I could for the good of the United States. When I met with Joseph Stalin (Ruler of Russia 1941-1953) I knew he was very cunning, but I never knew the extent of his sly cunningness until I came over here in the spirit world. Now, Stalin is over here too, but he lives on a lower different level of a spiritual dimension of life from where I now live. There are times when I can go down to his level and I can talk with him. We have had many conversations. He now is in a frightful condition. He needs love to enter his soul so that he will begin to realize the many wrongs he has done to his fellow men. It will take a very long time for Stalin to pay his debts to humanity. I also have debts to repay and I am doing my work to repay them."

* * *

GEORGE WASHINGTON, President 1789-1797.

On April 15, 1960, WASHINGTON spoke: "I am Washington, your First President. During my time the Presidency was not easy, though it was not operating on such a colossal scale as the Presidency now operates. But we had our difficulties in a different way. No President has an easy time of it. When a president takes office today he gives his life for that public office. It is very easy for presidents to make serious mistakes which affect the whole earth world.

"When I was President we had isolated ourselves from the rest of the world, but now the United States has come out of isolation.

"NOW ALL THE NATIONS OF THE ENTIRE EARTH WORLD ARE SCHEDULED TO BECOME UNITED TOGETHER AS 'ONE WORLD GOVERNMENT' WITH NO DIVISIONS. IT WILL BE 'ALL ONE UNITED WORLD.' "

On November 3, 1960, WASHINGTON said: "We have

built a table here in this yard at this center, and all the former Presidents are gathered around this table. We are not the Higher Master Teachers who are coming here from their other worlds. We are Human Beings the same as you are now, but we function in our spirit bodies. Our work at this time is of such a nature that we must come back to the earth and help with the political aspects of the United States. We are not working with the other Nations as this is our Nation where we are sent to help. We are diligently working in the United States of America."

ABRAHAM LINCOLN spoke on April 15, 1960. "We, who have lived and operated on the earth as former Presidents of the United States, will visit with you tonight. When I was in earth life I knew this 'truth' pertaining to the continuation of living in the spirit world after earth-life death. I heard the voices of my Spirit Teachers and knew who they were. I followed their instructions given through the psychic-medium who came to visit me at the White House. I was guarded and guided entirely through the instructions that my Spirit Teachers gave to me.

"Before I made my transition into spirit life, I knew I was going to lose my earth life, as I had been given a vision of it. When I made the transition, I stepped suddenly and quickly out of my physical body. It was like a great electric shock, as though I had taken hold of an electric dynamo, and with such force and such power that my whole physical body shook. And in that manner my spirit body stepped out of my physical body. I knew and was conscious of the fact that I had made the transition. There beside me stood all of my beautiful Spirit Teachers in their beautiful raiment. That was my proof that I was guarded and guided by High Raised Beings. Then they propelled me along with them, but I didn't know how, as I was not walking as I usually had walked. It seemed to me that I just moved, or floated.

"My Spirit Teachers took me to a very beautiful place where there were many hundreds of people. It was exactly like a great convention was going on, and I was to be the main speaker. When I entered the room a great ovation took place. And I thought: 'Is it possible that I, Abraham Lincoln, who was born in a log cabin on the earth, could have all this honor bestowed upon me?"

I had this glorified entrance into the spirit world, but I cannot put it into words how glorious it really was.

On this side of spirit life, I was then briefed during a period of time of what I was to do. I was told to go back to the United States to help my successors as they came into the Presidency. All of the former Presidents who now are in spirit life have formed themselves into a group to work together to help the mortal earth Presidents who preside in the White House."

PRESIDENT JOHN F. KENNEDY'S DEATH PREDICTED!

On June 20, 1961, JEHOVAH, an Extremely Highly Evolved MASTER BEING, gave the following predictions:

"Numerous earth people will be losing their earth lives in different ways and everything is changing. Even the elements, the heat and the cold, and all of the elements of the air are changing. The whole earth world is going through a decided change.

"I also say this to you that the Leaders of many nations will be removed, some by death and some by assassinations. And I say of the Leader (John F. Kennedy) who is at the Head of the United States, he will carry out only a certain portion of his rule. This too, will be entirely different from what the people of the United States expect. Everything will change and be different, and the reason for that is there will be great confusion all over the United States." (John F. Kennedy was shot at Dallas, Texas, November 22, 1963.)

* * *

ULYSSES S. GRANT, President 1869-1877

On April 6, 1962 GRANT said: "I have never been here before, but several of the other former presidents have come here at different times and have spoken to you.

"As I go back over the period of time when I was president, and view the periods of time that the later presidents have experienced, and compare their terms to the terms of my office, then I am glad I reigned at the time when I was president. During my time people lived more of a normal way of life. The extent of crimes were not as great as they are now happening on the earth.

"Your nation and the entire earth world are in the flux of great changes. I don't know how long it will be but every nation in the entire earth world will become a part of every other nation. The nations in the earth world will be divided in the same way as the States in The United States of America are divided.

"YOU ARE GOING TO HAVE A UNITED STATES OF THE WHOLE EARTH WORLD . . . A WORLD UNITED STATES OF THE WHOLE EARTH WORLD . . . A WORLD UNITED STATES!

"When all earth people can meet upon one firm foundation, where righteousness will prevail, then you will have Peace, Love and Harmony during this NEW AGE!"

* * *

JOHN F. KENNEDY, President 1961-1963.

On January 15, 1964, KENNEDY spoke: "A Great Master Teacher brought me to this center and said to me: 'You have held a high office recently upon the earth and I say to you, now you may speak.'

"I am now being educated on this side of life so that I can help to fulfill the office I so recently held as President of the United States. I can now work from where I now live. I know exactly how I must work for the good of all mankind, to help bring about peace on the earth.

"I am doing a much greater work here than I would have done had I remained in earth life. I am in Washington, D.C., most of the time, impressing Statesmen without them being conscious that I am there with them.

There are many men in your government who are greedy and deceitful. All of those who will not work hand-in-hand with the Spirit Beings will be removed from the earth."

COMMENT: Other Presidents who brought messages were, James Madison, Grover Cleveland and Woodrow Wilson. Lincoln spoke eight times and F.D. Roosevelt spoke five times.

PART 5

THOTH, THE ANCIENT ATLANTEAN
(Explains the Sinking of Atlantis)

HOW THE GREAT PYRAMID IN EGYPT WAS BUILT

ATLANTEANS FROM ANCIENT ATLANTIS
Predict Atlantis will Rise

GOLDEN TABLETS HIDDEN IN GREAT PYRAMID

A PSYCHIC TRIP INTO THE SPHINX

A GREAT CIVILIZATION BENEATH THE SAHARA DESERT

THE LOST CONTINENT OF LEMURIA TO RISE

THOTH, THE ANCIENT ATLANTEAN

A Great Being, THOTH, often came to our center and he gave us the following messages and teaching.

During the many long past ages of time, there have been many MIGHTY GREAT MASTERS . . . ILLUMINED WITH WISDOM, who were assigned to raise up all mankind on the earth.

The ISLAND OF ATLANTIS was a center of a Great Cultural Civilization. They were far advanced in the philosophic and the scientific sciences pertaining to the operation of the Natural Cosmic Laws of our Universe. They taught the basic sciences, such as geology, astronomy and mathematics. They had greater scientific knowledge that was far beyond today's known sciences.

The ancient Masters established Secret Mystery Schools that were places for teaching their philosophic and scientific sciences. Only qualified, carefully chosen students were admitted to those mystery schools. They accepted only those people who truly were seeking the higher, strange, hidden "truths" and the MYSTICAL SECRETS OF LIFE. The Mighty Masters always carefully guarded their wisdom teachings. They protected their ancient records in hidden archives, to make sure that their sacred records would not fall into the hands of profane people who would wrongfully and wickedly use the secret scientific knowledge contained in their records.

THOTH was born and raised on Atlantis. His father, THOTHMES was a Highly Evolved Priest-King. He was the Leader of the Priest-Kings who ruled the many great cities on Atlantis. Thoth was taught the ancient mysteries and he attained vast wisdom and mighty powers.

The highly educated Atlanteans had learned how to utilize many subtle, invisible forces and how to control and use the "ether" which exists everywhere in the universe. They used the powerful etheric energy existing within the atomic molecular structure of all physical matter. They had knowledge of the subtle, invisible, vibrating magnetic currents that flow throughout the universe and through all things, including Human Beings. Thus the Atlanteans had knowledge of the magnetic vibrations and their effects upon all people. The Atlanteans became highly skilled in controlling and using the universal magnetic currents of mighty power as a source of free radiant energy.

They built great atomic powered machines in which they concentrated and utilized the etheric atoms and the electronic beams of power, and the free flowing magnetic currents which they drew from out of the ethers in space—using those mighty cosmic-powered machines to neutralize the earth's gravity force. Such a machine enabled them to control and nullify the pull of gravity which holds material things to the earth's surface. Thereby, producing levitation of things, causing them to float in the air, and enabling them to hew, lift, float and transport from far distances, colossal, enormously heavy blocks of stones, such as massive pillars and the immense heavy stones used to build pyramids. They used those machines for many different purposes.

They built and traveled in vast, extremely swift mighty atomic powered space ships, which were only used by the Great Masters.

They knew the art of transmutation. Their alchemists could change the atomic structure of material things and transmute them into other forms, such as turning lead into gold. And their alchemists could activate and extract the spiritual quintessence of a healing herb, or of a tree, into an elixir having healing qualities.

They also knew how to make staffs containing powerful energy vibrations. The Evolved Atlanteans usually carried a staff with them. They used their staff to direct a ray of power with which to temporarily "stun" wicked people who were doing harm to other people. Their staffs were also used for other purposes. (Thoth and Hermes carried such staffs).

The Atlanteans, with their powerful atomic machines constructed hidden passageways leading far down deep for about 200 miles, to the inside of the earth. Those passageways connected Atlantis with the inner part of the earth. There in the inner sections of the earth different races of mortal earth-type people lived. (Some such people are still living in the inner earth during this period of time).

During the time when Atlantis existed highly evolved civilizations lived within the vast hollow sections of the inner earth. They were races of people such as mortals who live on the earth's surface have never seen. Only highly qualified Atlantean Masters were permitted to enter those carefully guarded passageways, and go deep down into the inner earth.

The Atlantean Masters knew that a GREAT PYRAMID OF MIGHTY FORCE AND POWER exists within the center of the earth. It stabilizes and balances the entire earth. And when that Pyramid of Power becomes even very slightly unbalanced, it causes earthquakes, tidal waves and other catastrophes to happen upon the surface of the earth. That Pyramid is very carefully guarded by Mighty Masters who have attained vast wisdom and mighty powers. They constantly control the pyramid's stability. They also guard it to prevent profane people, who might through some devious means, gain entry to a passageway leading to the pyramid. Such wicked people could, if they gained control of the pyramid, use it to accomplish their evil deeds, by throwing the pyramid out of balance and thereby shattering and destroying the entire earth.

The inner earth pyramid supplies all the light, power, energies and air necessary for the well-being of the inner earth's inhabitants. The inner earth also has rivers and lush vegetation of all kinds.

The righteous Priest-Kings ruled the various large cities on Atlantis. There also were many unrighteous people on Atlantis. Eventually, some of the highly educated men on Atlantis, who had been taught the great scientific secrets, became egotistical with their great knowledge and their souls fell into evil darkness. They became bombastic and brazenly displayed their knowledge of the hidden cosmic powers, using those mighty powers for destructive purposes. They had been taught about the hidden passageways to the inner earth and its Central Pyramid of Power.

The Priest-King Leader of Atlantis, with his vast wisdom, psychically knew the evil intentions of the fallen Light Brothers. He knew they intended to gain entrance to the inner earth and unbalance the Great Pyramid of force and power, which would destroy the entire earth. Therefore, the Priest-King Leader immediately advised all of the other Priest-Kings and the Guardians of the inner earth pyramid, that the evil ones were on their way down to the inner earth to accomplish their wicked deeds. And he instructed the Guardians of the Pyramid that only Atlantis was to be destroyed. The Pyramid Guardians knew exactly how to slightly tilt the Pyramid so that it would only destroy Atlantis, by sinking it in the Atlantic Ocean.

THOTH was advised that Atlantis was doomed to sink beneath the ocean waves. Immediately, he gathered together numerous secret records containing the far-advanced wisdom teachings. Also other things, such as great powerful scientific machines, and placed them into his vast spaceship. Then, he and his group of Highly Evolved Atlanteans, his Associates, men and women, entered his space ship and sped forth as Atlantis began sinking. They rapidly traveled to the land now known as Egypt. There, Thoth and his Associates established an Atlantean colony. In due time, they also migrated and also established other Atlantean colonies in various other parts of the earth's surface, where they built Pyramids and other massive stone structures in such lands now known as Mexico, Central America, and in South America.

Thoth had attained vast knowledge, far beyond our comprehension. With great electrical machines which he had brought with him, he built a passageway from Egypt to the deep inside of the earth, leading to the inner earth Pyramid. Then, he again used the machines with which to build the

GREAT PYRAMID in Gizeh, Egypt. He placed that Pyramid directly over the Central Pyramid of Force in the inner earth. Then he connected the Great Pyramid with the inner earth passageway.

When building the GREAT PYRAMID he used his electrical machines to draw powers from out of the etheric atoms existing in space, to hew the massive heavy stones at the stone quarry. There the stones for the Pyramid were their atomic forces to nullify gravity, and to change the atomic structure of the stones until they became lightweight. Then he floated them through the air into their exact positions. Then again, he changed the structure of the stones back to their original composition. The inner earth Pyramid of Force, over which the GREAT PYRAMID is built, helps to retard the deterioration of the GREAT PYRAMID.

The GREAT PYRAMID is the MIRACLE OF THE ANCIENT AGES. It incorporates the basic sciences of geology, astronomy and mathematics, and it is a masterpiece of engineering skill.

Inside of the GREAT PYRAMID, Thoth built hidden chambers and secret archives in which he placed the Ancient Atlantean Secret Records that he had brought with him from Atlantis. The records included Golden Tablets. Parchments and Scrolls, containing the ancient wisdom teachings. He also secreted the electrical machines and his spaceship in deep hidden places below the Great Pyramid.

The passageways that lead from the earth's surface to the inner earth are constantly guarded by Invisible Masters.

The GREAT PYRAMID was used as a mystery school where student adepts were initiated into the secret sciences. It also was used as a burial tomb for honored people.

(Thoth, the Great Atlantean, must not become confused with "Hojax" (an Egyptian) who named himself "Thothma" when he ruled Egypt. Thoth preceded Thothma by thousands of years, prior to the time that Thothma ruled Egypt. He also built a Pyramid).

HOW THE GREAT PYRAMID IN EGYPT WAS BUILT

On August 22, 1958, the GOD OF THE PYRAMID WORLDS said: "Mortal man on the earth has no idea of how the GREAT PYRAMID in Gizeh, Egypt, was built. It has been a secret how mortal people placed those massive stones that fitted perfectly, one on top of the other. I now will tell you the secret of how the Great Pyramid was built, because I was there when it was built by THOTH, an Ancient Atlantean.

"It was not built by the human power of mortal people. But it is true, there had to be mortal demonstration of bringing the colossal stones together at the Pyramid, guiding them into their proper places. Those massive stones were made 'light as a feather' for those mortals to handle.

"The Highly Evolved Immortal Masters in the spirit worlds understand electricity. They can control the elements of the universe and such things as magnetism, gravity and levitation.

"At the stone quarry, THOTH used ancient Atlantean electrical atomic-operated machines with which to draw powers from the atoms existing in space, to change the atomic structure of the stones and to nullify gravity, to eliminate the gravitational pull of force holding those colossal stones to the earth. Then the enormously heavy stones became weightless and were cut to their exact dimensions. Then they were floated through the air by the power of magnetism and levitation to the pyramid site, to their exact positions. Then again, the atomic structure of the stones was changed back to their original heavyweight composition.

"I now have brought to you the 'mystery' of the building of

the GREAT PYRAMID."

On January 16, 1959, a Great Master, named SPHERICAL, said: "The Atlanteans operated through a magnetic belt and they used magnetism and electricity for their of living and mode of travel. I now relay to you the 'secret' of the building of the Great Pyramid. It was all done through electricity and magnetism, and those massive stones were 'floated through the air.' Human hands only were used in placing them where they were supposed to go. They were floated into place.

"I have come to the earth to bring that same method to the earth people, which they will use when the New Age manifests."

ATLANTEANS FROM ANCIENT ATLANTIS PREDICTS ATLANTIS WILL RISE

August 26, 1960, during a session attended by Sara, her husband Ray and myself, Sara described Two Great Masters. "This is their first contact at this center. They are dressed in white and they wear peculiar cone-shaped headdresses. One Master is standing by you, Ray and Lara, the other Master is in front of you. He has in his right hand a tree. Its green short branches turn upward and its leaves are very shiny. Now, the one with the tree is facing me and his skin has a brownish-red color and he has reddish hair. His features are entirely different from any I have ever seen. They bring a new vibration and I can hardly open my eyes as they are so heavy."

Then the Master with the tree said: "We both have lived on the Island of Atlantis many thousands of years ago. I was a very important individual there.

"I have been instructed to come to this center. The tree I have in my hand, it was a symbolical emblem that was prominently used in the Island of Atlantis. It represents the symbol of spiritual growth. When a person has concluded and made the proper spiritual development, then he has the right to use the emblem of the tree. It was called the 'Tree of Life' and of 'Light.' Symbolical of understanding some of the ALMIGHTY CREATOR'S Vast Wisdom. Consequently, when we come here, or go to other centers, we bring with us the emblem that demonstrates our advancement. I will come another time and then I will bring to this center something that was very prominent on Atlantis. It was a Principle that was invoked and carried out on the Isle of Atlantis by those of us who were very highly evolved.

"We also had both good and evil operating on Atlantis, the same as what is now operating on the earth. The evil that was rampant on Atlantis was the cause and the means by which our land was demolished and submerged beneath the waves of the Atlantic Ocean. It's destruction was caused by people who had attained Great Spiritual Light and Wisdom and then used that knowledge for evil purposes.

"When I come again and take my place here at this center, I will bring to you, my child, Lara, information that comes from the ancient Atlantis. I now tell you this:

"THAT ISLAND OF ATLANTIS IS BENEATH THE OCEAN WAVES, AND THERE WILL COME A TIME WHEN THAT ISLAND WILL COME FORTH AGAIN . . . AND IT WILL RISE. THEN YOU WILL KNOW MORE OF ITS HISTORY.

"If this does not come about during your earth life, that has nothing to do with it, because whatever body you may be functioning in, wherever you may be, you will know about it. If you have passed from mortality to immortality, then you will know more about it than if you were still in mortality."

Then, the other Atlantean spoke to Ray: "My name is ALEXANDER, and I too have come from the Island of Atlantis. Now I ask both of you, Ray and Lara, to raise up both of your arms and hands, so that I can tie-in with your vibration and feel its quality. Now you may lower your arms again.

"We have come to the earth to bring information for the benefit of all the people on the earth. We are using you, the persons at this center as the 'dynamo' . . . as the mortal contact, which we need to make through mortals. This enables us to bring to the people on the earth greater knowledge that we have for them. The time has now come for us to begin our work. And again, I repeat what I have just said: 'You are the Power, or the Magnetism that we are using, so that we can tie-in and hold this Power, to bring earth people information about what is coming to the earth during this time, in one of it's evolvements.'

"There are many other Teachers who have instruments to come here and bring to you of their knowledge and wisdom.

Many thousands of years have passed since we, the Atlanteans, have lived in mortality on the earth. Since that time we have made great advancement in many other worlds. WE KNOW WHAT IS COMING TO THE EARTH WORLD!

"We have been far-advanced SAGES and MASTERS OF GREAT MAGNITUDE. That is why we have come, because only THOSE of that qualification, from now on, will come to this center and to other centers.

"The two of us will remain here until this session ends, because as it continues, we will get the substance and the feeling, and we will use your mind and your thoughts, as they go out into Infinity."

Tonight, September 16, 1960, as our session began, Sara said: "I see many MASTERS in white robes. This entire house is filled with them and even the yard. And as far as I can see in this area there are those white robed people. I also see them in the air and the whole earth world is surrounded with these GREAT BEINGS."

MELCHIZEDEK spoke: "I have my Light-Brother EPHRAIM with me, and EZEKIEL and ZECHARIAH also are here, and there are many others who have not yet introduced themselves. All are here for a definite purpose. They are not here to entertain you. They are here because they have a very definite task to perform and they must perform their tasks through mortals, such as yourselves, and especially through such a person as this channel, Sara. Whenever such a great assemblage of MASTERS occurs, it is because great issues are at stake all over the earth world. This is the way they facilitate their instructions and bring those things into manifestation that are necessary to be done.

"There always has to be a very definite reason why prophecies are given through this channel. Sometimes they do not take place for some time to come after they have been prophecied, but eventually those prophecies that were made here at this center do materialize."

Then ZARATHUSTRA, and ANOTHER MASTER showed themselves to Sara. Zarathustra said: "I have with me my Light-Brother, the ETHIOPIAN. We are doing something unseen that is sensitive. I come to you, my child, Lara, and the Ethiopian comes to the gentleman, Ray. We are doing something for the three of you that is far beyond your physical and your spiritual conditions.

"Something NEW is taking place in the earth and it will take some time to bring this work into focus."

Then the TWO ATLANTEANS again appear and show themselves to Sara. One of them said: "I invoke the Power and the Blessings upon this center. We have returned for a specific purpose. We are tying-in with the work of the 'Silence.' We have been assigned to this center."

Then Sara said: "As he was talking he built a light in here. It was a gold light and it enveloped both of them. The Atlantean who comes to Ray had a peculiar conical headdress, and the other one who comes to you, Lara, had a green tree."

October 6, 1960, Sara described what she saw in the room. "I see seven golden lights and now they have turned into the SEVEN GOLDEN AGE MASTERS. They have formed a circle and we are inside of that circle. The hands of each one of the Masters are very busy, doing something . . . making symbols with their fingers, using the ancient sign language. The writing of these symbols means something. I now will tell you what they are writing: 'Thou shall walk in the Light of holiness, for thou have been called to the earth from a Great High Pyramid World, by the CREATOR. And the Great Pyramid in Egypt is symbolic of that Pyramid World.' "

HERMES at one time lived on Atlantis. He is the most revered of the ancient Alchemists. He said: "In that Great Pyramid in Egypt, you will find secrets there that mortal earth man has never found to this day. There are certain archives in that pyramid that reflect and explain the writing that is in the SPHINX. And the lower passageway beneath the SPHINX connects up with the passageway of the Great Pyramid.

"I shall inform you of still greater mysteries and explain to you, that when the Isle of Atlantis existed, there were many underground passageways leading out from that Island. They were so deep down that mortal man living on top of the earth's surface, if he were to excavate, he could never reach down to those passageways.

"I also inform you, that down deep in the inner bowels of the earth, there are cities, governments, kingdoms and nations, that are operating in a similar manner as those operating on top of the earth. Those places consist of ancient civilizations that are entirely different from those now living on top of the earth's surface.

"I have now brought through the mind of the channel information she has never heard. This information also is brought to tell you that there are archives in the Sphinx and in the Great Pyramid. And there are hidden GOLDEN TABLETS and GOLDEN PARCHMENTS with hieroglyphic symbols written thereon. The day will come at some future time during this New Age, when people will discover, decipher and read those tablets and parchments.

"To bring into manifestation this New Age will be a gradual procedure, as the heavens and the earth gradually become interwoven together. Then all things will work together in Harmony, Peace and Love, when the proper time arrives."

On October 7, 1960, Sara explained: "I see TWO MASTERS in this room and one wears a conical pointed hat that is quite tall. He tells me he is HERMES. He holds something in his hand and he looks down at it, and it ticks, but it is not a watch."

Then the other Great Master said: "I have with me my very close Associate whose name is HERMES, and this is THOTH speaking. I am an Ancient Atlantean and I have recently come to this center during other sessions.

"The two of us are bringing a vibration here that you have not had before this time. And each time a new vibration is brought here it reflects a greater strength and a greater resiliency than what you have had before. By adding this new vibration to what is already established here brings a greater force.

"I say to you, my child, Lara, hold out your left arm and hand, extending it straight out into the vibration built up in this room. We are going to take a measurement and that is why we have this little instrument. It will reflect a measurement that is necessary to determine a certain distance in the Great Pyramid.

"We will bring you information of things that happened many centuries ago. You will be given the instructions and the enlightenment of what took place when the GREAT PYRAMID in Egypt was built. You also will be given a great revealment about the Isle of Atlantis.

"I say to you, there are great secrets of things that are hidden in the passageways, deep down below the SPHINX. Also, there are a number of mummified bodies placed in that symbolic stone statue, called the 'SPHINX.' Those mummies are the bodies of Great Men, and the earth people have no history of them. Some of them are mummies of people who had lived on Atlantis.

"THIS IS THOTH! I HAVE LIVED ON THE ISLE OF ATLANTIS!"

THOTH continued speaking and said: "I say to you, Lara, that you will be told, and you also will discover many things that are hidden, and you will reveal them in your writings to the people of the earth. Your writings will be very outstanding.

"I, THOTH, say to you that the nations on the earth world as a whole, will operate together as 'ONE NATION' in time to come. But it will take a long time before the nations will operate through 'One Principle.' It is a New Principle being brought to the people of the earth in this New Age. When that Principle begins to demonstrate, that is the time when the people on the earth and the Immortals from their spirit side of life, will be able to visit with each other and cohabit together."

Then HERMES spoke: "The Isle of Atlantis was in existence many thousands of years ago. It gradually is coming to the surface once more, but it will take many years before that happens.

"THE AQUARIAN GOLDEN AGE IS AN ESTABLISHED FACT."

(It has been reported that researchers have discovered massive pillars, building foundations and stone walls near the Islands of Bimini, the Bahamas and the Azores, in the

Atlantic Ocean. Those ancient ruins perhaps are part of the ancient Atlantis).

* * *

* This is a brief explanation pertaining to the conical pyramid shaped hats that many Great Masters wear. The hats contain cones of power that produce a great source of energy. The Masters use those cones of power for many important purposes.

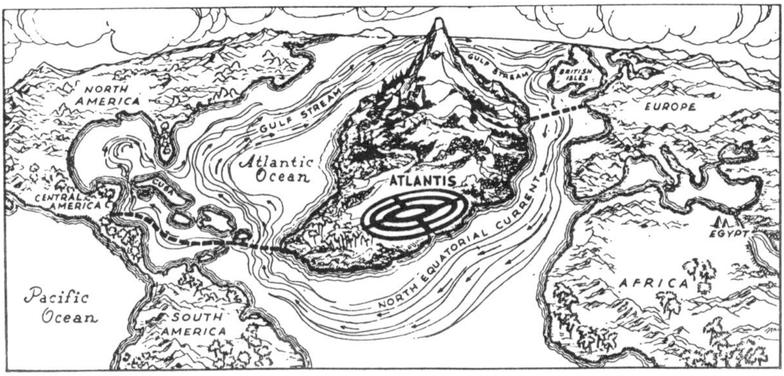

Map showing the locality of the ancient submerged island of Atlantis.

This very old map has been in my files about fifty years . . . I clipped it years ago from a newpaper that now is out of printing business. —**Laura Horst**

GOLDEN TABLETS HIDDEN IN THE GREAT PYRAMID

On April 20, 1961, the GOD OF THE PYRAMID WORLDS psychically took Sara and myself to the GREAT PYRAMID in Egypt. Sara described what she saw.

"We are now at the entrance of the Great Pyramid and we go down into a secret passageway. It is dark and musty. Then we enter into a large room. In that room are many odd looking things and some mummies of the ancient ages. There are hieroglyphic writings and pictures upon the walls. Each picture depicts a story of an event which took place centuries ago, pertaining to the earth life of those mummies.

"Now we walk through another narrow passageway and come into another room. This room seems to have a psychic light shining from somewhere. I see some stone benches and some stone urns and other stone containers of some kind. It looks as if this room was used for a meeting place. The walls are like hard rock and I see a man sitting on a stone bench. He is very ancient and wears a headdress like the Pharaohs wore and his long hair is hanging down. I don't know if he is alive or in what condition he shows himself. I see him sitting there very quietly as though there is no life in that body. I am wondering why he is shown to me.

"As I look up in that room there is a light shining down. There is a narrow ramp and we walk up on it and enter into a very spooky, narrow musty-smelling place.

"The God of the Pyramid Worlds is with us and he stoops over and pulls out two stones from the wall and it seems those stones are placed on top of each other. The lower stone is pushed in farther than the upper one which extends out. He mentally tells me that there are some 'Tablets' hidden behind those two stones. Now he tells me to reach my hand back there in the wall opening from which he had removed those two stones, and that opening is pitch dark. I take hold of something and it seems like a piece of metal. It is thin and when I pull that metal slab out there are hieraglyphics on it. IT IS A GOLDEN TABLET. He tells me there are many more of those Golden Tablets in that hole in the wall and they are standing on edge against a wall.

"The God of the Pyramid Worlds tells me that those Golden Tablets have been hidden there in that stone wall during many past centuries. No earth person knows those two stones can be pulled out. That was another different experience we had today."

Then, a Great Master, named the ETHIOPIAN, spoke through Sara: "When I was living in physicality, I was known as the Ethiopian. I was present when that Great Pyramid was built and I helped to build it. We used the power of levitation. Those priceless records, the Golden Tablets, with my own hands I placed them in that Pyramid.

"That Great Pyramid, like everything else, has an 'aura' around it which is its electrical vibration."

After the session ended Sara said to me: "If we could physically go into that Great Pyramid, I could lead you directly to where I saw that room with its two removable stones. Those two stones are about two inches thick and about eight inches wide. I don't know how long they are."

May 27, 1964, a Mighty Being, named "THE REDEEMER" said: "We have not spoken about the pyramids for a long time but the time is close at hand when you will find new information will be published about the Pyramids. Many of those Pyramid mysteries will be revealed and their 'Hidden Secrets' will be solved. Earth people will find parchments and scrolls containing "Spiritual Wisdom" which has been hidden since the Great Pyramid was built."

---

* In the *Los Angeles Times* of May 21, 1986 there was a news report of French scientists who, in making more accurate measurements in the interior of the Great Pyramid, have discovered hidden areas never before found, because the previous measurements were inaccurate. Scientists, at this time, are seeking ways to open these sealed areas.

A PSYCHIC TRIP INTO THE SPHINX

On July 10, 1959, during my session with Sara, MELCHIZEDEK took us on a psychic trip into the SPHINX. He spoke through Sara, saying: "We are walking into the opening of the Sphinx and it is cool in here. The sun never penetrates this stone image, yet we are not in complete darkness. The light from each of our souls is all that is necessary to show us the way.

"Now we have gone a long distance, through a tunnel, and we are going down seven steps. They are not moist and slippery, but they are cool to our feet. As we are bending over and looking down into the enclosure where we now are, we do not immediately see what is in it. But suddenly, we see a blue light in a blue atmosphere down there.

"After having gone down those seven steps we see a tomb. And as we look at that marble tomb, it is vividly clear and Sara is running her hand over a marble casket. There is a stream of light coming from overhead. We do not know where the light is coming from as there are no material lights of any kind. It is a light shining forth from the CREATOR, which has always shone its rays down upon this marble casket.

"Now the top of the casket has suddenly been removed and we see a mummified body lying in it. It is one of the ancient Pharaohs. Upon his eyes are two small golden discs that completely cover his eyes, and there are golden containers placed over both ears, completely hiding the ears. The features of this mummified form look like parchment, but there is a faint color in the cheeks, put there during its embalming to make him look lifelike. At the right side of the mummy there is a gold vitality rod and he holds its knob in his right hand. Now, suddenly, the lid of the casket has been placed back upon the casket.

"Now we come to the opposite side of this enclosure, which is just large enough to contain the casket and for three people to stand in front of it. Then, you Lara, turn and face the east, and you go up seven steps on that other side, but you could not reach that other side only through this sacred tomb with its casket containing the Pharaoh.

"We continue walking until we reach a certain place, and then we go up seven other steps. We find ourselves standing and looking at a large wall, upon which is something so surprising to us that it is almost taking our breath away as we look upon it.

"It is not a painting, but apppears to be a mural-type picture, with colors that are very bright and distinct and very beautiful. As we view that mural on that huge wall, we find it depicts certain events that happened during the earth life of that mummified Pharaoh who is in the marble casket.

"The mural pictures a chariot with two horses and a very young Pharaoh, who is flushed with excitement, as he is leaping out of the chariot. He holds a spear in his right hand which he had used to kill a lion. In the background of the picture it shows the lion lying on the ground.

"This mural faces both the east and the west, and as you look at it from one end to the other, there is a spirit light at the corner of each wall. It is a light that stands there and it has a flame. And that flame very frequently rises to the top of the enclosure and then it falls back down again until it lays out over the floor. Then it slowly creeps upward again, and when it finds its original place once more, then it rises up to the top of the ceiling.

"AT THAT TIME THERE IS A LOUD CLAP OF THUNDER AND THERE ARE STREAKS OF LIGHTNING IN THAT PLACE IN THE SPHINX. AND AS THE THUNDER CRASHES AND AS THE LIGHTNING FLASHES, WE FIND OURSELVES ON THE OUTSIDE OF THE SPHINX."

(We do not know how we were taken to the outside of that

tomb in the Sphinx.)

MELCHIZEDEK told us AKHENATEN was the name of the mummified Pharaoh in the casket.

Mortal earth historians tell about AKHENATEN'S earth life, and they spell his name in different ways. At birth, he was named Amenophis IV. Later, he changed his name to Akhenaten. This now, is a brief summary of his earth life, according to mortal earth records.

His father was Amenophis III. At an early age he appointed his young Prince son as his coregent, and for four years the two jointly ruled Egypt from their capital in Thebes. Then suddenly, an extraordinary change happened to the young Pharaoh-Prince. He was inspired to build for himself and his family an entirely new capital on a virgin site about two hundred miles south of Thebes. There, he and his beautiful wife, Nefertiti and their family established themselves.

It was revealed to them that the GREAT CENTRAL SPIRITUAL LIFE-GIVING SUN, which feeds all other smaller suns with its life-giving rays shining afar out over the entire universe, could be likened to an ALMIGHTY SUN-GOD CREATOR. Therefore, they worshipped the earth's sun as the new-found SUN-CREATOR, whom they named "ATEN." Then, in honor of his Sun-Creator, Amenophis IV gave himself a new name and became known as AKHEN-ATEN (or Akhenaton). During many past centuries the name "Aten" became "Aton" (or Atom).

AKHENATEN adopted that name after he broke with the Amen Priesthood of Thebes. He was a rebel against their established order of things. His ideas ran counter to all tradition. As a Co-Ruler Pharaoh, he put his ideas into practice. He cast out the Amen Priests who resented his numerous reforms.

In some devious way disgruntled persons poisoned him and he died in 1361-B.C.

THE SPHINX

The SPHINX is a stone statue located a short distance

from the Great Pyramid in Gizeh, Egypt. It is a mysterious riddle, symbolizing ancient wisdom. It faces east and it has unknown symbols and writings inscribed upon it. It has a lion's body and a human head.

On October 6, 1960, HERMES, a Great Master, said: "There are certain archives in the Great Pyramid that explain the writings on the Sphinx. And a lower passageway beneath the Sphinx connects up with the passageway of the Great Pyramid."

October 7, 1960, THOTH, an Ancient Atlantean said: "There are great secrets of things that are hidden in the passageways, deep down in below the SPHINX. Also, there are a number of mummified bodies placed in that symbolic stone statue called the Sphinx."

A GREAT CIVILIZATION BENEATH THE SAHARA DESERT TO BE DISCOVERED

December 12, 1963, a Great Master named SPIRIT BEING NO. 5, said: "I want to bring you history that has never been recorded in books on the earth. There will come a time when this New Golden Age manifests, when the Sahara Desert will bring forth 'Something' that the people on the earth have never seen and have never heard about. Far down deep beneath the earth's surface for many miles beneath the Sahara Desert, there is a civilization that is operating. Its people are not like people who live on the earth's surface. They are different types of Human Beings. They are far advanced in wisdom and understanding of life. They have advanced far beyond the advancement made by people living on the earth's surface. That civilization is not represented anywhere else on the earth.

"Their land that comprises their habitation is different from the land on top of the earth. The time will come when there will be an access to that civilization. There will be an entrance and a roadway made deep down into the earth, and eventually, that entrance will be opened to the people living on top of the earth. Then commerce will be interchanged.

"The Immortal Spirit Masters will bring to the earth mortals the secret knowledge of how to penetrate to that civilization that is living deep down beneath the desert sands. Then earth mortals will learn great wisdom from that civilization."

THE LOST CONTINENT OF LEMURIA TO RISE

February 20, 1959, a Mighty Master said: "I am known as the TWELVE POINTED STAR, and I am now taking the channel, Sara, on a psychic trip with me into my country that lies submerged under the waters of the Pacific Ocean and I am showing it to her.

"IT IS THE SUNKEN CONTINENT OF LEMURIA!

"Now we are submerged and are going down into a crevice that leads us down below the floor of the ocean, and we come to the sunken continent. We see many beautiful buildings as they have not been destroyed during the past many thousands of years.

"ONE DAY THAT CONTINENT OF LEMURIA WILL COME UP AGAIN, RISING TO THE SURFACE OF THE EARTH.

"I have come from far-off regions of the universe to bring to the people of the earth greater knowledge and wisdom about Life."

MAY 31, 1960, a Great Master named MIZPAH said:

"THE SUNKEN CONTINENT OF LEMURIA IS COMING UP OUT OF THE OCEAN!

"When that happens, there will be GREAT CATASTROPHES, and many people will lose their earth lives."

November 26, 1963, a Great Master named MICHAEL, said: "You have been told repeatedly that the earth world will change. That many great changes will take place upon the earth. There will be lands that are now occupied by cities that will disappear entirely. Then other sunken lands will reappear, such as ATLANTIS and LEMURIA. Then the true history of both of those long lost ancient civilizations will be revealed to people on the earth. The complete changing of the entire earth world will be different from what anyone can imagine."

(Many of the South Sea Islands, including Hawaii and Japan are remnants of Lemuria).

Outline map showing the names and divisions of the earth as used in this book. After the "Sinking of PAN." Map from Oahspe.

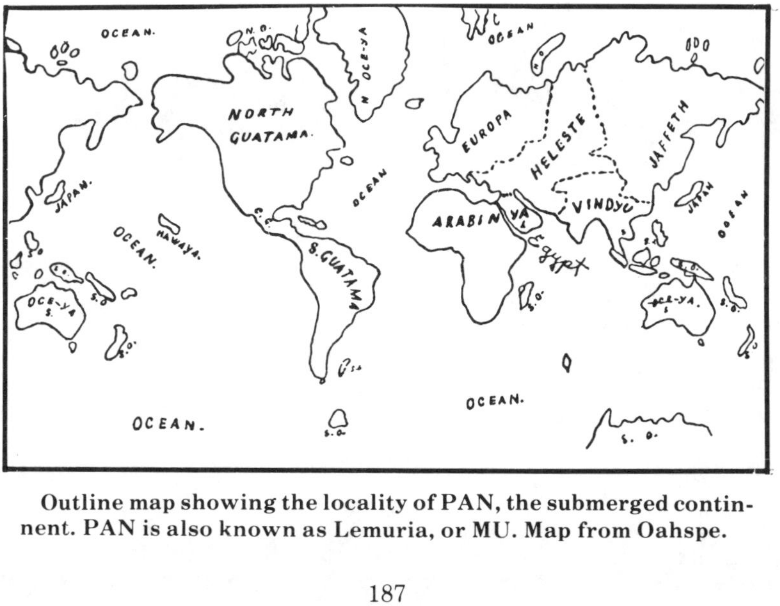

Outline map showing the locality of PAN, the submerged continent. PAN is also known as Lemuria, or MU. Map from Oahspe.

PART 6

DEATH OF A MORTAL EARTH PERSON

THE TAJ MAHAL LOVERS

CHAOTIC SPIRIT BEINGS ARE RESCUED

NERO, THE CRUEL ROMAN EMPEROR

DEATH OF A MORTAL EARTH PERSON

Each mortal person is a Dual-Being, having two separate co-existing bodies through which he functions during his earth-life. His heavy-water-flesh-body holds him to the earth. He has also an electronic-type spirit body that is warm and highly refined. But it is not a nebulous misty-body as it has substantial solidness. Its electronic elements are composed of electrical energies that function through much higher, faster frequency rates of vibrations than those elements composing his flesh-body.

Earth-man's spirit-body is so delicately refined that it can slip out of his flesh-body and then slip into it again just as easily as your hand slips into and out of a loose-fitting glove. His spirit-body can quickly pass through solid walls and other dense obstructions. It freely passes through the atoms of coarse physical matter in the same way as a man's flesh-body passes through water while swimming.

Man's flesh-body and his spirit-body are joined together by a most vital ethereal, electronic-type umbilical cable, called the silver-cord. This cord is a resilient current of throbbing life energy, and its elasticity is capable of extremely long ranges of extension, ranging for many thousands of miles. It permits Man's Own Spirit to leave his flesh-body and project "out" in his spirit-body to function separately. At all times the vital-cord keeps his spirit-body attached to his flesh-body.

EVERY HUMAN BEING IS AN ELECTRICAL SPIRIT BEING!

Death of the flesh-body occurs when the linking-cord is severed. Man's flesh-body then is discarded by his own personal Spirit in the same way as mortals discard a worn-out heavy garment. Then man's Own Spirit becomes "free" and he continues living in his own spirit-body, which he clothes by the power of his own thoughts. Then he can freely travel about on the earth's surface and in the lowest astral spirit plane that is joined onto the earth's surface. At the time of his earth death he retains his same personality and his same mentality and he remains exactly the same person. However, he may find himself to be in a dazed, confused condition and does not know what has happened to him.

During earth-life when families have had loving relationships, then the transition into spirit-life of a loved one may be a great heartfelt loss to those who remain in earth-life.

Many earth people do not know that their departed loved ones anxiously await the arrival of their beloved family members to join them in spirit life. They joyously greet their loved ones when they enter spirit life. They lovingly care for them and teach them their new ways of right living in the spirit world.

The Beings in the spirit world gradually progress from the lower spirit planes to the higher planes.

The many millions of Spirit Beings dwelling in the different graded spirit planes above the earth are not idle. Many of them volunteer to perform certain work to help other Spirit Beings who are in need of assistance.

Those volunteers organize themselves into separate rescue groups, called Spirit Missionaries. Such groups usually are composed of family members and their friends. They work together to rescue chaotic, confused, dazed Spirits, who they find wandering about on the earth and in the lowest astral spirit planes. Such rescue groups also rush immediately to help earth people who suddenly lose their earth life due to tragic accidents and great catastrophies of various kinds.

On October 17, 1958, a Spirit Lady brought her husband

into our session room. Then he spoke through Sara: "While I lived on the earth, I was a very lazy, selfish person, living entirely for myself, which caused great hardships for my large family. My wife always was kind to me and tried to help me overcome my selfishness. She died many years ago before I came over here into spirit life in the year 1957. Now we cannot stay together because I cannot live and breathe in the higher spiritual vibrations where she lives. I can only stay there for a little while and then I choke up because it is difficult for me to breathe. I then hurry back to the lower vibrations where I can freely breathe. I now must work to undo all the wrong things that I have done in the past, so I must work with other people who need help. The Teachers tell me that I will gradually rise higher until I can live and breathe in higher vibrations with my wife."

During the many sessions when the Master Teachers brought messages to us they spoke about different topics, such as our universe, life and death of a mortal person, etc.

They explained that the heavenly spirit world surrounding the earth is a counterpart of the earth. It is joined onto and orbits with the earth. It is a duplicate, a replica of the earth's surface. It has its spirit mountains and oceans and all such things as those existing on the earth's surface. That spirit world is divided into several graded dimensions, or levels, or planes. Each plane has its own graded degree of spirit light. The higher the plane, the brighter the light. There is no darkness in those planes. The very lowest plane that exists close to the earth's surface is called the astral plane. It does not have very much light because it has a murky, blue haze, like a dense fog. There are Spirit Kingdoms located directly above each country on the earth's surface. When earth people make their transition into spirit life they enter into the Spirit Kingdom that exists directly above the country from which the mortal departs from the earth.

* * *

ADDENDUM:

January 24, 1963, RAMESES II stated: "The ENTIRE

UNIVERSAL COSMOS IS A VAST OCEAN OF ELECTRICITY."

* * *

The LIFE FORCE in every human person is electrical. Each person lives forever throughout all the long ages of eternity.

THE TAJ MAHAL LOVERS

The SHAH JEHAN was the Mogul Emperor of Hindustan, now known as India. He was born in 1592 and died in 1666. His favorite wife was Mumtaz-I-Mahal and he made her his Empress. She was very beautiful and highly evolved spiritually. He idolized and worshipped her. She died at the age of 44 years.

The Shah deeply loved Mumtaz and they enjoyed the ultimate in wedded bliss. She died while giving birth to their fourteenth child.

The Shah became heartbroken and he devoted the remainder of his reign and his vast wealth in the building of a tomb in her memory, which he named the TAJ MAHAL. It is the world's most famous and most beautiful mausoleum. Its entire structure is of white marble and it has four slim minerets which flank the beautiful great dome. Light enters inside the building through lacy marble screens. The walls of its chambers are decorated with flowers portrayed by dazzling precious gems inlaid in the white marble.

The Taj Mahal, at Agra, India, has immortalized one of the most tender of all love stories. And this now, tells the rest of their love story about what happened to them after they made their transition into spirit life.

April 11, 1958, A Great Powerful Spirit Being spoke at our center. He said: "I want to explain to you, Lara, what happens to Sara when we use her as a channel through whom we speak. During that time she feels like she is under a mild anesthetic. She feels like she is half-out and half-in her physical body. And in that condition she knows what is taking place.

"Today, we are taking her completely 'out' for a short time, to test her to see if she can perform an important demonstration. This is a test that she must take to see if she can face the ordeal, and to see if she will have any fear of the unknown. We ask you, Lara, to sit quietly during that testing time."

Then there was a long time of silence, until finally, the Great One spoke through Sara, explaining the test given to her. "We took her out mentally into the astral spiritual realm, through the 'fog of her mind' and we gave her a great searchlight to use as she wandered about, believing she was all alone in that deep astral fog. She kept calling out, 'WHERE AM I?' But only an echo responded to her calls. Then we brought her back from her astral flight and she now is in her normal condition. She stood the test, as she had no fear, only faith, knowing that not one thing can harm her. She does not at this time know what a great service she has performed today for two Spirit Beings, during the time she was in the deep astral fog."

Then Sara, herself said to me: "My, that was an ordeal, let no one ever tell you that Spirit Beings can't take you over. You are as helpless as a baby and you can't move or think. And your head feels like it is in a vice and you have no energy power. That really was an experience and it wasn't very pleasant. I never thought this would happen today."

On April 18, 1958, the SHAH JEHAN spoke through Sara, telling how he was brought into contact with his beloved wife, Mumtaz.

"I am the Shah Jehan, and during the time that I lived upon the earth with my favorite dearly beloved wife, I lived entirely for her, to make her happy. Then suddenly, she was taken from me. And to honor her, I built the Taj Mahal. Then after I made my transition from earth life into spirit life, I have been wandering all over the earth world and in the earth's astral spirit regions, seeking for my beloved one. I had no peace in my soul.

"Then a Great Spirit Master came to me and said: 'You seek peace, so come with me and I will show you the road that will lead you to your loved one.' Then I found myself beside this woman (Sara) and I said: 'Why have you brought me here, as I do not know this woman.' And the Great One replied: 'No, you don't know her, but she too has suffered heartaches and she knows the heartaches you have been enduring. Now, with this great magnetic belt that has been placed around her, you will be touched on your forehead. Now, open your eyes and look about you, and as you look through that great spirit window, I will let you see your long-lost beloved one and she will see you.'

"Then, my long-lost loved one said: 'Ah, yes, my beloved, I see you.' I asked her: 'Why do you not come to me and why do you stand so afar off from me, while I stand here with outstretched arms'? She replied: 'Oh, heart of my heart, and of my soul, it is needful that you lift your eyes to the CREATOR, and that you do not worship me, but only worship HIM. Your worshipful attitude where I am concerned, has brought you to your present confused, lost condition. You must not worship a former mortal being but only worship your CREATOR. You have made an idol of me and now you must undo that. You must forget your own personal self, and you must work to seek out the many souls that are lost, even as you have been lost during such a long time. As you seek the lost souls and put forth your efforts to help enlighten their burdens then you will be compensated. And as you spiritually grow higher and ever higher, in your understanding of life, then you will once again be able to join me. Then we can become as 'One' forever."

Then the Great One spoke to the Shah: "I say to you, this woman (Sara) to whom you have been brought by us, little does she know what she has done for Two Loving Souls. Your beloved one, for whom you have been searching, has been spiritually living through higher spiritual rates of vibrations, on one side of the spiritual veil separating both of you. And you have been functioning through much lower rates of spiritual vibrations, on the other side of that spiritual veil. Those different spiritual rates of vibrations have kept you separated from each other. You could not meet each other without the help of a mortal earth person, such as Sara."

Then the Shah Jehan said: "I now am mentally taking

this woman (Sara) with me, to show her the Taj Mahal that I built for the other half of me, who was my Soul-Mate. I want her to see what one person could do for someone they loved dearer than life itself. I am going to take her into that beautiful Taj Mahal building. And now, I say to her: 'Come with me, and you will see the reflected Spirit of she that I loved so dearly. Now look into that mirror, and you will see the reflection of the one that I idolized. See the great beauty of that Soul. She was half-angel and half-physical, and the great joyous love that I felt in my soul when we were together was without bounds.' "

Then again, the GREAT ONE said: "Only a few earth people who have become qualified, can be used by Spirit Masters to release bound souls. Those who wander about in the earth's low astral spiritual regions, seeking for each other. I say to the Shah: 'We used Sara to help us release your bound Soul, and to bring about this meeting with your loved one. And as you progress in your spiritual advancement, you will gradually rise up to that higher level of life where your loved one dwells. Then it will be made possible that both of you can join hands and go forth together, as you journey through your spiritual progression of eternal life.' "

CAMACITO, A Highly Evolved Spirit Being said: "I want to explain to you, Lara, why this channel (Sara) was used to accomplish the bringing together of two historical Souls, who had been separated in the spirit realms during a very long time.

"The numerous Great Ones who come to this center, have been chemicalizing, magnetizing and sensitizing Sara's physical body, which has become a dynamo storehouse of those different powers and energies. We draw and use those as they are needed by us to help us to perform our work on the earth and in the astral realms.

"Those two Souls could not be brought together, until there could be found a mortal Being on the earth, from whom we could draw and use the specific atoms, chemicals and energies we need, stored abundantly in a physical body. That is why Sara was used to accomplish the first stages of reunion of those two loving Souls. There are many millions of Spirit

Souls wandering about in the low astral heavenly regions. They are searching for their loved ones, for members of their families and friends, and some of them have become earthbound. This happened in the astral outer-space regions near the earth's surface. All of the astral regions must be cleansed of all chaotic wandering Spirits. All earthbound Spirits must be removed. Then they must be taken to their proper places and taught righteousness. This must be done before this NEW AGE can be brought into full manifestation." I said: "Thank you, Camacito for this explanation."

Then Sara said to me: "That was a funny experience I had today. When they took me to India, at first, I wandered about and I was in a strange country. I saw many women and they smiled and talked with each other. Then all at once I saw the Taj Mahal. And before I could enter it, the Shah Jehan appeared and he took me into the Taj Mahal and showed me the reflected Spirit of his beloved wife."

CHAOTIC SPIRIT BEINGS ARE RESCUED

On October 17, 1958, my two brothers, Hank and Herman, reached and brought a spirit man named JONATHAN, into our session room. Hank said: "Something has happened to this Spirit and he is dazed and confused. We have done everything that we can do to help this young man. It seems that he has had some kind of a shock and he has not been able to come out of it. That is why we have brought him here today. We kindly ask the Teachers here to help him."

Then TES, the great Healer at our center, said: "This young man received a shock when he fell from an airplane and he still feels that he is falling. I say to you, my sons, now take him over to the spirit building that has the letter "X" and you put him into that building. It contains many different kinds of spirit machines that have fluids in them. One machine has a certain fluid that this man's spirit-body is lacking. The Spirit Doctors in that building will operate on him and put some of that special fluid into his spirit-body so that it will strengthen him. I will come over there to help revive him so that he will feel normal again. Perhaps next week he will be able to speak to you."

On October 24, 1958, my two brothers again brought JONATHAN into our session room. Hank said: "Today, Jonathan is going to speak." Then Jonathan said: "I still feel somewhat confused. I remember I was a strong, healthy Human Being and I was sitting in my airplane seat watching the bright clouds. Then suddenly, there was nothing but black darkness. Then I wandered about in that darkness and I didn't know where I was during a very long time. Finally, when I opened my eyes I saw a light. Faces were bending over me but I didn't know them. Then I heard someone talking to me, telling me 'You are not dead.' Then I said: 'Of course I know I am not dead and you don't have to tell me that. I just want to know what is the matter with me.' " Then Hank spoke again: "We are glad that he now is able to put his thoughts into speech. This is a new beginning for him but he still is very weak and at times he 'blacks-out' but we will bring him in again when he is more fully recovered." Then they departed with Jonathan.

Sara said to me: "Lara, I have taken on that young man's condition and I am so 'light-headed' that I can hardly sit up. I am so weak that I can hardly speak these words."

On November 1, 1958, my two brothers again brought Jonathan into our session room. Herman said: "We now have been able to bring Jonathan into the 'light of understanding.' We have been teaching him like a child. Now finally, he has come to the point where he is able to grasp what we have been telling him. Now I say to him: 'Go ahead and speak.' "

Jonathan said: "From the time that I sat in the airplane, until this present time, what has taken place is a 'blank' where my mind is concerned. But my friends, the brothers, have told me of things that have happened to me. These friends have convinced me that I have died from my earth-life in an airplane accident. I know that somewhere I have a mother over here where I now am, who died long before I did. These friends have promised me that they would take whatever you have here, and I don't know what it is, but they call it some kind of power. They are going to use that power to find my mother for me." Then he loudly shouted: "OH HERE SHE IS . . . MY MOTHER!" Then his mother who was sobbing with great joy, said: "OH, BLESS YOU, JONATHAN."

Then Hank said: "Now we will stand aside and leave them together. It was the power from this center that brought his mother here. This is a happy reunion and we too, know about such a reunion as we have experienced that."

Sara then said to me: "When Jonathan was speaking and asking for his mother the chills went up and down my back and when I turned my head I saw his mother and saw them rush into each other's arms, but I didn't see her face. Now she has found her son and that is wonderful."

One of Sara's friends named Elizabeth, during her earth-life was a trumpet medium. She died January 4, 1944, at the age of 76 years. She was the next Speaker: "Now you know that these things happen constantly in the world of spirit. Families are brought together and reunited. That is the work assigned to me, to help Immortal Spirits to find each other. To bring families together. Now this mother and her son will be sent to a school, because the mother had no knowledge about the afterlife in the spirit world when she made her transition. Bringing her here was an awakening for her. This center is well known by numerous Spirit Teachers for this kind of work."

JONATHAN, on November 14, 1958, spoke: "I have come in on my own power as I have received great strength and I have learned much. I am not yet fully recovered from the shock of finding my blessed mother. It was all made possible through this center.

"As I look at my own spirit body, I do not see it any different from the body I had while I was on the earth, and I am absolutely no different, insofar as I can recognize myself, and I think as I always did. I see the world of earth and I see this other world and they seem to be linked with each other and I do not know where to separate them. On my own volition, I have come to the earth world and I see the earth as I have always known it and I see the place where I have lived. I also see the world in which I now am living. They tell me it is the spirit world. Unless earth people have been taught the principles about life and death, and then when the time comes when they make their transition, they do not know how to make the separation between the earth world and the spirit world. And they need the help of the Spirit Forces to teach them the principles, as your brothers have taught them to me. I am now here to express my thanks to all those people who have helped me. To thank the Spirit Teachers, this medium, and you, Lara, and your two brothers for all the help that has been given to me. God Bless You."

Then Hank spoke: "The last time when we brought Jonathan in here and he said he wanted to find his mother, in that minute when he expressed his thought, there was his mother beside him, and their reunion was very touching. I say to you, Lara, our family members who are over here, we are glad to be here. It is worth all our efforts to help persons who have made their transitions and had become dazed and confused, not knowing what had happened to them. Those are daily happenings. We draw power from this center to help us do our work. Mankind knows very little about the different facets of life."

TES then said: "Now you see that we demonstrate our healing work in the visible earth and also in the invisible, in the spirit world surrounding the earth. We do this healing work to help our fellow men. The magnetism and the vibrational healing power going forth from this center can heal earth people that are many thousands of miles distant from here. We are here to help all those people who are ready and willing to receive our help. Many people have been healed. These are the arts, in our humble capacity, that we are bringing to the people on the earth."

On February 26, 1960, the morning newspapers reported:

"At Rio de Janiero, a big U.S.A. Navy Plane crashed over Sugar Loaf Mountain, killing 61 persons."

"At Shannon, Ireland, an Italian airliner, with 52 persons aboard, exploded, killing 27 persons and the others were injured."

During our afternoon session on February 26, 1960, TES said: "I have asked that a double-portion of spirit power flow through here in this room, in order for us to accomplish what has been assigned to us to do today. To rescue and heal all of those people who have lost their earth lives yesterday in two different plane accidents. Many of those passengers still huddled together near the crashed planes are confused and do not know what to do and they do not know where they are as things seem so strange to them in the spirit world. Some of those persons who died have been rescued by their Spirit Loved Ones who have taken them away with them. The others who have had no help will get their help today, as we are making the conditions whereby the Spirit Healers will rescue and heal them. The Spirit Rescue Groups have been very busy rescuing people all over the earth world who have made their transition into spirit life."

April 10, 1963, the atomic submarine THRESHER, built by the United States, sank today in the Atlantic Ocean, with 129 men aboard. Then, during our session today, TES and his close assistant named "ELL" and their other many Spirit Healers, took the healing power from our center and they rushed to the rescue and aid of all the men who perished in the Thresher submarine disaster.

On November 9, 1959, a GREAT BEING said: "During your earth life you operate through your own 'Spark of Life' which is your 'MIND, THE CREATIVE POWER.' And with your POWERFUL MIND you can reach far out and travel to many far distant worlds. According to the advancement that your 'Mind-Spark' has developed, that is how much you can see and hear in a distant world. But when your earth life is finished, then you operate in another rate of vibration, in another order of life. However, your own 'Life-Mind-Spark' continues operating in the same way, with your 'Creative Thoughts' flowing through your POWERFUL CREATIVE MIND. Our work in spirituality is brought into created manifestation through the Creative Power of our Thoughts. We can have a thought and quickly that thought will fully manifest and operate regardless of how small or how large the project is."

NERO, THE CRUEL ROMAN EMPEROR

Nero was Caesar Lucius Domitius, born year 37 A.D. He ruled from 54 A.D. to 68 A.D. He lived in a golden palace.

On December 19, 1960, NERO brought us this message telling about his earth life when he ruled the vast Roman Empire. He was a very cruel ruler.

First, Sara was shown a vision, she said: "I see the Roman Colosseum as it appeared when the Roman Emperors held their sporting events. There are hundreds of spectators. The Roman Charioteers are driving their chariots very fast around and around before the large crowd of people. The drivers are yelling and whipping their horses to make them go faster. I get the name of NERO."

Then NERO spoke: "The scene viewed by the channel was the way in which we celebrated events during the time when I ruled the Roman Empire. We celebrated lavishly with great banquets of food and drinking intoxicating liquids, for our personal enjoyment. Little thought was given to our brothermen who had little or no food with which to sustain themselves. We, in the high offices thought only of our own self and we lived for what honor we could gain for ourselves. We had no feelings of compassion for our people. We did not consider their wants, needs or sufferings. In fact, we, the high rulers were the very ones who inflicted great pain and suffering upon our people, especially upon those who called themselves 'Christians' who had drawn away from the Roman beliefs.

"Oh, I say to you, that I wish I could live my earth life over again. I have wished that many hundreds of times. I have wished that I could have known this great truth about the after-life, the spirit-life.

"Oh, how I regret that I killed the Christians in the lions' dens, making sport of their death by the lions before thousands of spectators who also reveled in seeing the Christians devoured by the lions. Those spectators had lust in their hearts as they too did not know this great truth, that we must repay our debts and our cruel acts that we have done to our brother-men.

"I speak now to all of my subjects, to all the people who lived in the Roman Empire during my rule, saying to them:

" 'All of you Spirit Souls who are within reach of the sound of my voice, hear me as I speak to you now, with compassion and love and with pity in my heart. You, who have suffered due to my cruelties which I inflicted upon you, I ask you to listen to me once again.

"Oh beloved ones, little do you realize the mental sufferings and tortures I have endured, due to my cruelties to you.

"Beloved Ones, with all the love of my Soul and heart, I reach out to you now and ask your forgiveness. I vow to you that I shall continue to work with might and main until I have redeemed myself with each one of you. All of you dear people who are still in the darkness of your soul, still living in that same vibration of pain and fear and have been suffering during all of these many past years, ever since you were tortured and deprived of your earth life due to my lustful gain. I vow to help you to bring yourself out of your misery and I will help to teach you the greater truths of life, which I have learned since my transition into spirit life.

"I ask you to now come forth and let me help you and to redeem myself in your sight and in the sight of my CREATOR. I, and all of those many other people who are working with me will help you. It is my duty to help you as I must right all the wrongs that I inflicted upon you. I ask you, in all the humility of my soul, to forgive me, even as my FATHER-CREATOR has forgiven me. I will work hard to repay my indebtedness to you who I have harmed and placed into darkness. I send out my love to all of you. I, the former Great Emperor NERO of Rome, am now humble before thee.' "

PART 7

ASTOUNDING! . . . STUNNING! . . . NEW AGE PREDICTIONS

EARTH WORLD IS BREAKING UP

ALL NATIONS WILL HAVE UNUSUAL DISASTERS

ALL WAR WEAPONS TO BE NEUTRALIZED

SPACE TRAVEL & SPACE STATIONS

CELESTIAL TELEVISION MACHINES TO BE BUILT

NEW AGE THOUGHT MACHINES TO BE BUILT

NEW INVENTIONS WILL PREVENT AUTO COLLISIONS

JUNGLES TO BE CLEARED OUT

THERE WILL BE BUILDINGS ON THE OCEANS

NEW AGE FOOD

FLESH BODIES WILL RECEIVE ARTIFICIAL IMPLANTS

FLESH BODIES OF EARTH PEOPLE WILL BECOME CHANGED

EARTH WORLD IS BREAKING UP

On August 29, 1960, MICHAEL, a Great Master, showed Sara a psychic vision, which she described: "I see a FABULOUS BIG GOLDEN LIGHT and it shines all over the earth. And now again, I see a very beautiful Spirit Being, and he seems to be hanging in space. He is robed in all white. He says he is MICHAEL, THE ANGEL OF LIGHT. He is holding a large globe of the earth world, and I see streaks of lightning coming out of it, and the globe is turning and turning. He still is hanging there in space. He says this vision is significant of SOMETHING that is coming to the earth, of GREAT MAGNITUDE.

"Those streaks of lightning are coming out of that globe in every direction. Now, he breaks the globe of the earth wide open, just like you would break an egg. He took his two thumbs and broke that globe. As I look inside of that globe, it has a very unpleasant odor, like rotten eggs, and there are many people inside of that globe.

"Among that great crowd of people inside of that broken globe, there is One Person who is very outstanding. He gives me the name of IMMANUEL, and I see him very plainly. He is a very beautiful Being and wears a long white robe that has many folds that trail back of him. The people in that broken globe, some of them have raised their arms, and others are on their knees praying for help. I do not know what this signifies. Now, out of the sky, around that Beautiful Being, I again see flashes of lightning and I hear thunder. Something very strange and important is taking place. I have never been given this vision before this time."

IMMANUEL said: "I now quote an ancient prediction

that was given in ancient scriptures: 'And in that day, shall MICHAEL, THE ANGEL OF LIGHT stand up. He is also a CHRIST! When he stands up then the Great Quakes will commence. And it was as though the earth was broken into pieces.'

"That prediction was foretold ages ago, before people on earth had a bible. That information was given to earth people during the time when Atlantis and Lemuria were in existence on the earth. A great portion of the information that is recorded in your Holy Bibles, was originally taken from the hidden, ancient parchments.

"I, IMMANUEL, say to you, this vision shown to this medium today is to alert earth people so they will know that the fulfillment of the predictions given in the ancient scriptures is near at hand.

"I have come to the earth at this time, because we are now in the beginning of the breaking-up of the earth world. This is not a destruction that will come all at one time, but it will come about at different times. This breaking that I have reference to will cover a period of many years. The people on the earth are standing on a 'Precipice' without being aware of what is happening.

"As the earth is being cleansed you will see destruction of all kinds, and famine and deaths of millions of people. You may wonder why these destructive things are permitted, and people made to suffer. I explain to you, their suffering is but of short duration. There is just a sudden shock in which the mortal's own Spirit, in his spirit body, comes forth from out of his physical body. Then, his Spirit feels no more pain, and his lifeless physical body is discarded in the same way as you would discard a worn-out heavy coat.

"The earth has arrived at the time when ALL THINGS are to be made NEW, and the old order of things must be destroyed. You cannot build a new foundation upon an old one. When I refer to the old foundation, I refer to the various governments in the countries on the earth. Each country at this time has a Great Master, who is a Highly Evolved Being, who has been assigned to each particular country. Each Great One is standing at the helm, conducting the procedure in the way it is to be conducted. They are standing ready, waiting for the proper time when it arrives.

"The whole earth must be cleansed of all its iniquities, before the NEW HEAVEN and the NEW EARTH will come into full development during this NEW AGE."

ALL NATIONS WILL HAVE UNUSUAL DISASTERS

On January 4, 1958, the BRIGHT MORNING STAR BEING showed Sara a psychic vision of an extremely large flame of fire. He explained that the vision is symbolical of what will happen to Russia. He said:

"THE WHOLE NATION OF RUSSIA WILL HAVE ITS BAPTISM OF FIRE."*

On July 17, 1959, CONFUCIUS spoke through Sara. He said: "There is a great change ahead of the whole world. Especially great changes will take place in the communistic countries of Russia, Red China and Cuba. They will have shattering holocausts and blood baths. Russia and Red China are 'Gog' and 'Magog.' They will be overthrown and communism will fail.

"I also say to you, that the 'Vatican' that sits on seven hills, will come down, like a house of cards.

"During these catastrophes many people will be removed from the earth. There will be a 'Sign in the sky' when these shatterings happen. Then, at that time, millions of righteous Spirit Beings, who are hovering over the earth in their

---

* CONFIRMATION OF PREDICTIONS:

On April 29th, 1986, Gorbachev released the news of a devastating nuclear reactor accident in CHERNOBYL, near KIEV, U.S.S.R. Attempts to keep this secret were in vain, as radiation fallout spread to and was detected by all neighboring countries. Headlines: WORST DISASTER SINCE HIROSHIMA.

At present writing as this book goes to press (May 23, 1986) the holocaust fire appears to be finally under control. Unbiased facts are hard to come by, but the radio news is that thousands are still being evacuated every day.

spaceships, will gather up into their ships, all those people who lose their earth lives. Their Own Spirits will be taken to heavenly kingdoms prepared for them.

"At the end of those holocausts, all unrighteous nations will become harmonious nations."

During a session on September 15, 1961, a Mighty Being, named the ILLUMINATED ONE, showed Sara a psychic vision which she described. "I see a winter scene in the Ural Mountains of Russia. It is very cold there, and I see some Russians who wear fur caps with lapels over their ears. They are working on a secret missile that they have built and have kept it hidden in the Ural Mountains where there is snow. That missile is in the shape of a large bullet. Now, they have put it upon a very long sledge, with its point upwards. They are looking up into a very gray sky."

The ILLUMINATED ONE spoke: "I am showing Sara what the Russians are planning to do. THERE IS NOT ONE NATION ON THE EARTH THAT CAN DO ANYTHING IN SECRET!

"I have psychically taken Sara to that place, down deep inside of the Ural Mountains. There is a great excavation, and I am taking her in there. I say to you: 'All the Imps of Hell' (so to speak), are in that large underground cave.

"For many miles that mountain has been excavated and it contains many large buildings. They have all sorts of machinery of many descriptions, and destructive weapons, such as long range missiles. Some of the people working there have been given life sentences. The Russians know they will never be able to leave that mountain to reveal their secrets of what is taking place there.

"The Russian Leaders intend to enslave all the people living on the earth. They have a diabolical plan to take over and rule the entire earth. They have established a condition of such magnitude of destruction, because they are entirely controlled by the evil Spirit Beings that are so rampant upon the earth and in the earth's lower heavens.

"When the Russians detonate their atomic bombs, they do not know that the winds of the atmosphere can shift, and their radioactive fallouts will come right back upon them.

"The Russians' secret unrighteous activities are well known by the Great Masters who have come to the earth. Those Masters have the power to change the vibrations in the air and change the winds of the earth. The Rulers of Russia will destroy many of their own people, due to their own destructiveness. Because of Russia's diabolical plans, they have built this very secret atomic missile base.

"The CREATOR'S patience is far-reaching, so HE has allowed them to do extensive excavations in those Ural Mountains, until they have nothing left but a shell. And, when the proper time arrives, THERE WILL BE AN EXPLOSION THERE OF SUCH MAGNITUDE, THAT EVERYTHING WILL EXPLODE AND BLOW-UP . . . AND THEIR OWN COUNTRY WILL BE LAID IN ASHES. THUS, THE NATION OF RUSSIA WILL HAVE ITS BAPTISM OF FIRE!

"By inspiration of the righteous Spirit Beings, the ALMIGHTY CREATOR has sent his mandate to the Rulers of Russia, to change their destructive ways."

On January 18, 1964, MICHAEL, the Great Being, said: "The United States of America will not be excluded from the drastic changes that are coming to the earth, but the United States will not be destroyed.

"Russia and Red China will experience all the evil things they have inflicted upon mankind on the earth.

Drastic changes will come about on the earth when people least expect them to happen.

"Those drastic changes will consist of catastrophes of great magnitude and they will take place all over the earth world."

(Russia completely controls their news media. However, reliable news sources have reported that a major, terrible atomic explosion happened in the Ural Mountains in 1958).

ALL WAR WEAPONS TO BE NEUTRALIZED

SPHERICAL, on February 27, 1959, said: "It is very necessary for you people here at this center to know very definitely WHAT is taking place on the earth.

"THERE WILL NOT BE ANOTHER GLOBAL WAR!

"The numerous Great Masters who have come in their spaceships to the earth, will not allow the atomic power of the destructive weapons to operate. That power will be cut off and the instruments of war will be neutralized. Every one of those weapons of war will not be able to operate and leave the ground to display destruction. However, those missiles that are built to go into outer space for constructive purposes, they will operate perfectly. Neither Russia or the United States, or any other nation, will be able to destroy each other with the destructive weapons of war.

"Those weapons and missiles that people on earth have built, will be made for entirely different purposes. They are the vanguards, the replicas of the spaceships that the earth people will build to travel to different worlds and other solar systems.

"The earth has come into a NEW AGE CYCLE and the firm establishment of the NEW EARTH and the NEW HEAVEN is taking place.

"This is outstanding information that has been given to you, and it is TRUTH in such a manner that no one can contradict it. Little do you realize what I am telling you."

On July 26, 1962, MELCHIZEDEK said: "Now you may have wondered about all of those destructive things of war which each nation is building to destroy each other. That will never happen, and I will tell you why. We, the Great

Ones, have enough power so that those instruments can be made impotent. And, if nations want to use them for destruction, then nothing would happen, because they would not function properly. The destructive power would have been taken from them. Those implements of war will never function for destructive purposes, but they may be used for constructive purposes.

"THERE IS NOT ONE THING THAT IS IMPOSSIBLE WITH THE CREATOR."

SPACE TRAVEL AND SPACE STATIONS

January 10, 1958, a Great Master, named CAMACITO, said: "I have come to the earth for the express purpose to build in the earth's vibration the magnetic electricity power in the atmosphere. Also, to build and increase the magnetic power of the bodies of mortals. The magnetic power controls the chemicals in a mortal's body. The Masters will draw the magnetic electrical vibrational power from earth people to help them do their work on the earth. They draw that power from mortals in the same way as if they had plugged into an electric light socket. They need that magnetic power from mortals to enhance their own mighty powers.

"We, the Great Ones, are going to demonstrate 'Something' on the earth that will be very vital and important.

"A great number of spaceships will come to the earth. Many are now patrolling the earth's atmosphere. Numerous people will see them and talk to those persons in the spaceships. Some mortals will enter those spaceships and go to other planets."

January 16, 1959, SPHERICAL said: "Many far-off solar systems and their numerous worlds, are conducted entirely through electricity and magnetic currents of power, existing everywhere in space. Mortal earth man has but a faint idea of what electricity can do.

"YOU ARE COMING INTO THE NEW ELECTRONIC ELECTRICAL AGE.

"That is why at this time of the earth's evolvement, electricity will play a very important part.

"Many of the great spaceships, or conveyances, that are coming to the earth, have a magnetic belt surrounding the ship. That magnetic belt of power is magnetized and electrified, so that the spaceship can be propelled and brought safely to its destination without an accident. That electrically illuminated belt of power enables earth mortals to see the spaceship. However, before a mortal person can come near, or can enter one of those ships, the electrical power surrounding the vehicle must be turned off. Because that great force of power is of such great strength that it could fatally injure mortals.

"In the future, space scientists on the earth will build and send spaceships and satellites of great magnitude to other worlds. They will place many SPACE STATIONS, SPACE PLATFORMS and SATELLITES into outer space, and they will build SPACE CITIES where people will live.

"All worlds and all such things as spaceships, space stations and satellites, are held in space by the ALMIGHTY POWER of the SUPREME CREATOR, who created the Universal Cosmos.

The Highly Evolved Great Masters use various kinds, types and sizes of space vehicles when they travel to other worlds.

"The Mighty Masters and their millions of Highly Evolved Spirit Associates are coming to the earth in Great Colossal Space Vehicles. In the future earth mortals also will build colossal space ships and massive space stations. Great spacecrafts will become as common and as numerous as your airplanes are today. In about from twenty to thirty years these things will begin to demonstrate. Spaceships will 'shuttle' back and forth from the earth to other worlds.

"If it would be impossible that mortals could not travel to other far-off worlds at some future time, then we, the more farther advanced Immortals, would not be bringing this information to your understanding. There would be no reason for us to come to this center and talk about them.

"The time will come when mortals will live and travel in their own spirit-body, while their flesh-body remains alive on the earth. When space travel becomes prominent, earth people will freely leave their flesh-body and travel forth in their projected, electronic-type spirit-body. Many mortal adepts understand this technique and are doing this spirit-body projection at this time.

"It is necessary to explain how this spirit-body traveling will be done. Each family on the earth will have a special 'Lying-In Room' in their home. They can go into that room and securely lock the door, then lie down on a bed, where their physical body will remain undisturbed until they return to it. Deliberately, they will easily project their Own Spirit in its spirit-body, out of their physical flesh-body, and travel forth in their spirit-body to a designated space station. There, they will board a space ship and travel forth to other worlds."

March 7, 1960, Sara was shown a psychic vision of a GREAT HUGE TALL TOWER, and a semi-oblong spaceship. Then psychically, she was taken into the inside of that spaceship into a round room, which she described: "The walls are made of shiny smooth wood, such as I have never seen on the earth. There are seats made of that same kind of wood and they are placed all around that circular room, and the room has a window. Now, as I sit there, the spaceship begins to travel, spinning very rapidly. I do not know where it is going."

Then, MELCHIZEDEK said: "You have been told that you now are living in a New Age, and that everything on the earth is changing and being made new. The channel has been shown, on a miniature scale, how people will travel in the not too far distant future.

"That EXTREMELY COLOSSAL HIGH TOWER that was shown to her, will be built on the earth at some time in the future. It will be much taller than you can now imagine in your mind. There has not yet ever been a building built on the earth equal to what I am trying to describe to you. That HUGE TOWER will be used for space travel to other worlds.

"Before space travelers go to the top of that tower, for a proposed space flight, they first will go to offices on the ground floor, where provisions are made for passage on space flights. This will be done in the same way as people now arrange for transportation at airports.

"People will not go to the very top of that tower by the way of stairs. There will be a new machine that will be built for that purpose. Earth people know nothing about that machine at this time. It is not an elevator. You will go into a cage, or a carriage conveyance on the ground floor, and it will be lifted to the top of the tower, through the use of a certain power. When you get to the top of that tower, that is the place where the passengers will wait until the spaceships arrive. The ships will come to their designated platforms which are enclosed.

"There will be a connection between the tower and the landing places of those spaceships. Atomic power will propel those spaceships. The atomic power that earth scientists are now delving into, is the same power that has not yet been perfected for general use.

"The different space platforms located at the top of that Huge Tower will be called 'Ports of Call.' Many spaceships will arrive and depart from them at the same time for other worlds. There are many billions of worlds in the CREATOR'S vast cosmos. To reach all those countless worlds takes all the ages of eternity.

"When a mortal person, in his spirit-body steps out of a spaceship, upon a different world, he will look exactly as he appeared when he left the earth.

"When Sara said the spaceship she had entered was spinning so rapidly, that was done to show her the rapidity of speed through which those spaceships will be manipulated.

"Now, I will enlarge on this and enlighten you so that you will understand under what circumstances you will travel to other worlds. If you arrive at a Space Tower in your flesh-body, you will not be permitted to travel in that flesh-body. At that tower, there will be provided special 'Lying-In Rooms' where earth mortals will lay down their physical body on a bed. Then, his Own Spirit steps out in his spirit-body, and it will look exactly the same as his flesh-body. His spirit-body will be clothed with the proper garments. This is accomplished by a mortal's Own Spirit, by its use of Thought Power. That is how mortals will be prepared for their space flights to other different worlds. Upon their return to the earth, they will reenter their physical body.

"This system of spaceship traveling is already in operation in many other solar systems. This information is new and strange for people living upon the earth. But it is necessary that this information now be given to earth mortals, because the earth has entered this New Golden Age.

"The spaceships and other things found on many other worlds, will not be built of earthly kinds of substances, but of other types of substantial substances.

"Such substances are as real to a man's Own Spirit when it operates in his spirit-body, through his spiritual senses, as material earth substances are real to his physical senses, when operating through his flesh-body.

"Remember, YOUR MIND IS THE REAL YOU, regardless of what kind of body you are wearing, and regardless of where you may go, or what you may do."

CELESTIAL TELEVISION MACHINES TO BE BUILT

On December 27, 1957, the MOON GOD said: "Spirit Beings now are working with a group of physical engineers on the earth, who are trying to invent an etheric television machine. Such a machine has already been invented in the spirit realms of life. When that machine is fully developed in materiality and when it begins to demonstrate it will open up astonishing revealments. It will show the millions of far-off other worlds and their inhabitants. Earth people will see many unbelievable, astonishing things. Such a television machine will change the religious teachings of all the orthodox churches.

"When that time arrives, Lara, your records will reveal that you received a great number of messages from Highly Evolved Beings. Their message foretold the inventions of FANTASTIC NEW AGE MACHINES."

August 14, 1959, SETHANTES said: "The events that took place thousands of years ago, those things are still in the atmosphere, in space. They are still going on.

"There will come a time when different machines will be built on the earth. Your televisions, at this time, are very crude, but they are a beginning. The future New Age television machines will be so attuned that mortals will be able to see and hear things that took place on the earth during the many past ages of time. Mortals will view great battles that were fought, and see the individuals, just the same as though they were mortals today. You have no idea of the things that took place on the earth and in other worlds in the universe in the past dim ages!

"In the future, people on earth will build huge things, and

they will have large viewing screens in those buildings. There will be no previous preparations made for the display of what will be shown on those screens.

"All past events that have happened in the universe are recorded in the etheric vibrations existing in the vast cosmos. Even today, these words now being spoken through Sara, will remain forever in the vibration of the universe. Words in the ether surrounding us will be heard, and past events will be seen on the screens of the New Age Machines. These are only some of the future wonders that are coming to the earth."

NEW AGE THOUGHT MACHINES TO BE BUILT

On September 12, 1958, a Great Master, named THE UNKNOWN ONE, said: "There will be a great machine built that will be greater than any machine that now has been invented. This new machine is already in the vibrations to the extent that the earth scientists will begin to think about it.

"Eventually, there will be a machine built that will be the means by which earth people can reach out into the universe. They will be able to reach far back to the beginning of time, and pick out the THOUGHTS and the DESIRES of mankind living at that time. Those vibrations will be relayed back to the earth. They will appear on scrolls, which will in language translations express the THOUGHTS, IDEAS and DESIRES of ancient people, and they will be recorded on that machine. Mortals on earth will hear the very voices of people who vocally expressed those things. This is what will be known as SOMETHING SO ASTOUNDING, that mortal man, during all the past ages of the earth, has never seen or heard anything like it.

"Also, earth people will hear the sounds of the peoples' voices from the Ancient Isle of Atlantis, as that great holocaust and that great flood, took place when Atlantis sank in the Atlantic Ocean. You will not be able to understand their language. Histories of ancient civilizations will then be translated and told to the people on the earth.

"We are working diligently for this NEW GOLDEN AGE to fully manifest when everything will be made NEW. There will be no death in mortality during that time of this NEW AGE. That will be the time when the earth's heavens and the earth will have become blended together as the same place.

"People on the earth have never before received such fabulous wisdom and demonstrations as those that are now coming to the earth."

December 15, 1962, THE UNKNOWN ONE said: "There now is a NEW MACHINE . . . A THOUGHT MACHINE, in the process of development. When this machine becomes available, then through that object, mortals will hear of things they never have heard, nor have ever imagined. Because it has not been feasible or desirable until now for mortals to know of these things.

When the THOUGHT MACHINE begins to operate, it will be the most FABULOUS THING that man's MIND has ever invented. Not only will you be able to carry it in your vest pocket, as it will be a small object, but you also will be able to talk to far distant worlds. You then can call to your loved ones and friends, who already are in immortality, calling them in the same way as you now call your friends on the telephone. It will be just as simple and just as easy as that. You might be sitting in a conveyance, or in your own home, and be THINKING "I want to talk to my mother, or other loved ones." Then, all you need to do is to send out your THOUGHTS, and your Thoughts will become connected with this little machine. IT WILL BE DONE BY THOUGHT.

Your THOUGHTS alone will be responsible for the information you will receive. You will not think aloud, as you will silently "JUST THINK" and it will be your THOUGHTS. The conversation will be carried on in THOUGHT where you are concerned. You will hear the conversation and believe they are speaking loudly. As the message is conveyed to you by thought, no one else will hear it. This can take place in a room with many people. They too, with their own thought machine can make connections with their friends in immortality, and no one else will hear the *message* they receive. This is advanced knowledge I am bringing to you.

Also, there will be ANOTHER DIFFERENT TYPE OF A THOUGHT MACHINE, that will be used to detect and pick up the THOUGHTS of other people. When such type of a THOUGHT MACHINE begins to operate, then no person can start destruction. Because as soon as they have a destructive thought in their mind, just that quickly, someone will pick up the destructive thought of that mortal person. There will be no secrets, as all things will be revealed. When using such a machine, no matter what a person thinks, it will be revealed to his fellow men. Their thoughts will be in the form of colored waves of vibrations. When another mortal person sees those waves he will be able to offset those destructive thoughts.

"In the future, during this New Age, mortal man's mind will be revealed to his fellow men, just the same as it now is revealed to Human Beings in other worlds. They can read each other's thoughts. Everything is laid bare. In the millions of other worlds, where Great Spirit Beings abide, they do not care if their thoughts are known, because they are thoughts of love and of helpfulness, and thoughts of constructiveness. On the earth world . . . there is where secret thoughts of hate, greed and destruction thrive.

"Now you may ask, 'Why do these destructive things happen on the earth?' You must realize the earth is where mortals get their experiences with the evil as well as with the good. How else could they tell right from wrong?

"I shall bring you much knowledge pertaining to the New Age. You know very little of what this New Golden Age consists of, or realize just what it signifies.

"In the future I shall give you information, such as you have never heard before. You have read about the Ancient Isle of Atlantis, but I say to you, there were other civilizations dating farther back than Atlantis. Other civilizations will be revealed that have existed on the earth. No mortal person knows the exact age of the earth. There have been many civilizations on the earth during the past millions of years that have never been recorded by earth mortals in their history records. When this New Age fully manifests then all things will be revealed.

"Those thought machines that I have been talking about, when they are perfected, you will be able to contact those long past civilizations that existed millions of years ago. Then you can ask them how they lived and conducted their lives. All of that information will be made known and it will become history.

"You can realize, as your mind reaches out, while I am speaking to you, Lara, what an interesting life people will have, through this new knowledge and information that will be brought to mortals on the earth during this New Age. Then you will understand much that seems unbelievable now.

"Mortal earth men will tie-in with Human Beings dwelling on other worlds. Then mortals will visit back and forth with those other worlds."

NEW INVENTION WILL PREVENT AUTO COLLISIONS

January 2, 1959, ENOCH a Great Master, said: "There will be a NEW INVENTION built that will be the means to avoid auto collisions. It will be a "gadget" that will stop the car instantly, regardless of the speed it is operating under at the time.

"This is what we, the Masters, from our spirit side of life, want to bring into manifestation upon the earth. For instance, if a person would suddenly dart out in front of the auto, then the gadget would be 'quicker than the eye' and instantly, it would stop the car. Also, suppose two cars were ready to collide, being too close to each other, then this gadget would stop the collision. That magnetic-powered gadget would instantly stop that other car.

February 6, 1959, THE UNKNOWN ONE also elaborated on this gadget. He said: "In the future there will be a New Invention that will instantly and automatically stop automobiles. It will be a 'magnetic' invention, and be brought about through atomic power that exists everywhere in the universe as 'free radiant energy.'

"There will be some changes made in the automobiles of the future. Their mechanisms will be changed. The auto manufacturers will build an auto which will be a revolution in auto manufacturing, compared to the autos they are now building. When they build this new type of car, there will be this new invention built into it. This magnetic gadget will be placed upon the dashboard panel of the car, where the driving instruments are installed. It will have the magnetic power so that when the auto comes too close to a danger zone it will automatically stop the car.

"This invention is still far-off in the future wonders that are coming to the earth."

(May 25, 1972, Newspaper Headline stated: "RADAR IN CAR CUTS COLLISIONS." In New York, an experimental automobile Radar System has been designed to prevent rear-end "tailgaiting" collisions. This was announced today by RCA. The radar mounted on the front of the car, tracks the car ahead on the highway, or in heavy traffic, recording distance apart and speed. When the separation between the two cars is too small for the speed, the radar flashes a light and sounds a warning buzzer for the driver).

JUNGLES TO BE CLEARED OUT

February 22, 1960, Sara is shown a symbolic psychic vision of a jungle, which she described: "I see a woman there, who has black hair and big-looped earrings. She has bare feet and is doing a very peculiar dance. I now am in that jungle where everything has grown so profusely and the trees and vines are entangled. I see many monkeys and I hear a lot of squawkings of all the different kinds of birds and other wildlife. Now, I see a stream of running water and I am walking in deep grass, all alone. I don't know if I am in a swamp, but I am trying to get to that water. As I get to that water, there is a great big crocodile lying there along the edge of the water. I also see a boat coming down that river. It looks like a wooden-box type of a boat and there is a wheel in front of it. The boat is coming towards me, and as it comes along the shore, I see many dark people on that boat. Somehow, I am now on the other side of that river water and I see nice farm land. I don't know what this vision signifies."

Then ZARATHUSTRA explained: "The medium, Sara, was psychically taken to that jungle, in her spirit body, and she was shown what will take place when the jungles are cleared out and turned into farming land.

"The natives living in the jungles will become civilized. Their minds will be spiritually inspired and their 'Thinking' changed. They will be educated and then live like other normal Human Beings."

THERE WILL BE BUILDINGS ON THE OCEANS

February 17, 1961, The Great Being named UNKNOWN ONE, said: "In the future, there will be buildings of many different kinds, shapes and sizes, built on top of the oceans. All to be built there for a special purpose. They will be floated, like a ballast, and be balanced and stabilized, but not fastened to the earth. They will not submerge, and a storm will not be able to tip them over as they will float along with the waves and the tides. Nothing will destroy them. Inside of those buildings will be new discoveries that earth man today has never heard about. They will tie in with the space age knowledge.

"The oceans will be mined, and researchers will find many unheard of minerals and ocean plants that will be used for food during this New Age."

(News Report dated May 5, 1979, stated:

"PACIFIC UNDERWATER VOLCANOES DISCOVERED"

Active underwater volcanoes spewing rich deposits of silver, copper and other metals and minerals, have been discovered deep down under the Pacific Ocean. The volcanic vents are described as being "lined up like a row of factory chimneys, belching black smoke." They are located in an area in the Mexican waters off the Gulf of California).

NEW AGE FOOD

April 3, 1959, SETHANTES showed Sara a psychic vision of a desert. She saw an irrigation canal flowing through that desert from one end to the other. Then the desert became fertile land with trees, lush green gardens, lakes and houses.

Then SETHANTES explained: "In the future, the deserts will be used to grow many products that have never before been grown on the earth. As you well know, there are deserts in the United States, such as the Mojave desert. Also, there are other deserts in various parts of the earth, such as the Great Sahara Desert and the Gobi Desert.

"The United States will start reclaiming one of its deserts with irrigation canals. Then, special seeds, containing New Age seed atoms, that have never before been planted on the earth, will be planted to grow seedlings. Then, those seedlings will be distributed and used for all other deserts in the earth world.

"The seedling plants will be lush and green and will contain all the necessary elements and requisites needed for the maintenance and good health of human bodies. Those New Age food plants will need to be planted only once, because they will grow so prolific. And, as the seasons come and go, they will, of their own accord, re-seed themselves, and yield their great abundant supply of nutritious food. There will be no need for cattle and other animals to be used as a source of human food.

"These New Age food plants can not be planted until the proper time of this New Age Cycle arrives. Then the prolific abundant yield of this New Age food, will produce enough food in all the deserts to supply all people on the earth.

"THEN TRULY, THE DESERTS WILL BLOOM LIKE ROSES."

September 22, 1961, A Great Being named NEBULA, said: "In the future, chemicals will be found in a certain mountain, which will be used to prolong earth life."

FLESH BODIES WILL RECEIVE ARTIFICIAL TRANSPLANTS

On September 28, 1960, TES, A Great Healer at our center, said: "Now, we speak about the physical conditions of pains and sufferings by earth people.

"YOU WILL FIND THE PHYSICAL BODIES OF PEOPLE WILL BE RENEWED. Organs will be replaced in the human flesh body. They will be artificial organs, not made of flesh and blood. Researchers will find a substance that will answer the same purpose and do the same functions. Such as replacing a flesh and blood heart, stomach, kidneys, liver, or whatever may be needed to repair a body. This substitute substance has not yet been discovered, but it will be discovered. Then, you will find that mortal people will renew their flesh bodies, and they will live much longer on the earth."

THE ABOVE PREDICTION HAS BEEN VERIFIED

February 24, 1981, a news report stated: "New artificial heart will improve our lives. First human implant of an artificial heart will be performed by Dr. William DeVries, of the University of Utah Medical Center. Dr. William Kolff, famed for inventing the artificial kidney machine, is the pioneer inventor of artificial organ implants. He heads the Medical Organs Program of the University of Utah Medical Center. Dr. Robert Jarvik joined Dr. Kolff's research team. He is the brilliant inventor of the artificial heart.

"The mechanical artificial heart is handmade of polyurethane plastic, marine aluminum, nylon, artificial fibers, graphite lubricants and high-strength alloys."

On December 2, 1982, at the University of Utah Medical

Center, Dr. William DeVries performed the first artificial heart implant on Dr. Barney B. Clark, a retired 61 year old dentist, from Seattle, Washington. Dr. Clark lived 112 days with his artificial heart. He died March 24, 1983.

FLESH BODIES OF EARTH PEOPLE WILL BECOME CHANGED

April 8, 1959, A Very Powerful Being named "TRIUNE GOD" said: "I want to tell you why we, the Great Ones, have come from our highly advanced Triune Worlds to the earth. It is because mortal people must be built up physically, mentally and spiritually in such a manner that they will be able to perform their necessary work in a similar way that we are performing our work. So that when the Spiritual Government . . . the New Kingdom, when it is firmly established on the earth, all people will become perfectly balanced and be able to properly operate on the earth. All things must be perfectly balanced.

"We are Great Beings, having great knowledge. We come to the earth to bring our vibrations and our knowledge for the purpose of preparing humans for what will take place on the earth when the proper time arrives.

"People on the earth will tie in with other people living on the planets in the earth's solar system. And also tie in with people living on other worlds.

"The Triune Mighty Masters have come to the earth to give mortal people greater power . . . to strengthen their bodies and make them more powerful. To renew their bodies, so that physicality no longer will be weak, so that diseases cannot attack their bodies.

"The physical and the spiritual bodies of mortals will be changed, strengthened and renewed, so they will have greater vitality and power. All of the spiritual centers in the bodies of mortals must be opened-up. These changes are vitally necessary, so that people will be able to conform and live in their new way of life that is ahead of them in the future, during this New Age."

August 15, 1958, the STAR BEING said: "Magnetism and electricity compose life. Mortal earth people have very little magnetism and electricity in their flesh bodies. A flesh body mainly functions through the bloodstream. The spirit bodies through which the Immortal Human Spirit Beings function, are composed of electricity and magnetism. They travel by the power of their thoughts."

July 1, 1964, A PURPLE RAY BEING said: "I am going to acquaint you with the fact, and explain to you, that the NEW ELECTRONICS now coming through the universe, are affecting all people on the earth. Also, affecting all life regardless of what each phase of life consists of.

"ALL LIFE IS NOW BEING ELECTRIFIED WITH HIGHER ELECTRONIC VIBRATIONS. The atmosphere itself is producing a great overflowing of electronic particles. If you could see them, you would see them falling down upon the earth like snowflakes in the winter time, but they are particles . . . *electricity*. They are invisible to mortal eyes, but they have form and shape, each one resembling a tiny white disc of light. You cannot feel those particles because they have no solid substance, but they do appear as little discs of light, having a certain power, now manifesting in Human Beings.

"I am speaking about these particles manifesting in the HUMAN FLESH BODY, and I DO NOT refer to them as manifesting in man's other spirit bodies. The flesh body is going to be transformed from flesh and blood, to a BODY OF LIGHT-SUBSTANCE, comparable to the fleecy clouds in the sky, or you may liken it to a sunbeam. It takes in all of these propensities. Yet, the body will look exactly as it does now, but it will be in its refined condition, functioning through a higher, faster rate of vibration.

"This is advanced information, because this body-transformation will gradually take place, unnoticed by the people. And it will change just as rapidly as the 'MIND' of the individual changes. If the individual puts forth his greatest efforts to desire to attain to greater spiritual wisdom and understanding of life, then that is how these changes take place.

"When the physical body has been transformed to the extent that I have mentioned, there is no need to make a transition (dying) as mortals now do. You will be able to function in that transformed body in both the physical earth world and in the spirit world, because you would be functioning through a higher, finer rate of vibration, enabling you to function in both dimensions. YOU WOULD BE ABLE TO MAKE YOUR TRANSFORMED PHYSICAL BODY INVISIBLE TO THE PHYSICAL EYE OF MORTALS, such as who are now living on the earth in their present flesh bodies. YOU WOULD MERELY BE CHANGING YOUR OWN RATE OF VIBRATION, TRANSFORMING THE COARSER VIBRATION INTO A HIGHER RATE OF VIBRATION. Then you can expand your powerful mind."

FUTURE RACE OF EARTH PEOPLE WILL BE OF ONE COLOR

February 21, 1958, A Great Master named "PO" said: "The time is now here for the people on the earth to become of 'One Mind,' of 'One Creed' and 'One Color.' That is what will take place when the right time arrives. Then, the people of all nations will be of 'One Color.' And there are no people now on the earth who have that new shade of color of skin, that humanity will have when this New Age fully manifests. There will be no pure whites, blacks, yellows, or red-skinned people. There will be only one color.

"New Age chemicals now are raining down upon the earth, and mortals are breathing-in those chemicals, and they are eating foods grown and chemicalized by those new chemicals that are falling upon the earth. Those chemicals will change the chemicalization of the flesh bodies of earth people, thereby changing their skin color. Those changes generally will become more prominent as people absorb those new chemicals. The new, unborn future generations of mortal people, will be the ones who will have this new skin color, because their parents will have absorbed great amounts of those new chemicals.

"I cannot describe the exact shade of skin color that people will have . . . but it will be more like an "IVORY COLOR."

Also, by the twinkling-of-an-eye, mankind on earth will become changed in their ways of 'Thinking' . . . because their 'Minds' will become changed. The reason why this will happen in this New Age, is because of the new chemicals that are now raining down upon the earth."

DESTINY OF THE UNITED STATES OF AMERICA

October 7, 1960, The Great Being, ROCK OF AGES, said: "The whole United States Government will change and work on different principles. There will be great changes made in the way the United States Government now operates.

"THE DESTINY OF THE UNITED STATES WAS FOREORDAINED TO BE USED AS A 'BEACON LIGHT' FOR THE REST OF THE WORLD. Its DESTINY is to bring the Light of Knowledge and of Spiritual Truth to the people on the earth. Symbolically, it represents the 'LIGHT THAT GOES FORTH FROM THE CREATOR' and reaches into all nations.

"THE NORTH AMERICAN CONTINENT WILL NEVER BE DESTROYED BY A GLOBAL WAR.

"There will not be a global war fought on the soil of the United States. But you may have drastic destructions, such as cyclones, earthquakes, tidal waves and other great disasters, at different times in various places. This information will stand you in good stead WHEN DRASTIC HOLOCAUSTS HAPPEN THROUGHOUT THE WORLD."

On August 3, 1962, a very beautiful, highly evolved Immortal Female Being, said: "I am the LADY OF DESTINY. I have been assigned to the United States of America. I have come to the earth to bring about many changes in your prominent government offices.

"Many mortal females now will begin to be elected to your high government offices. They will help to bring about many necessary changes. The women will help to bring into manifestation THIS GLORIOUS GOLDEN SPIRITUAL AGE."

THERE WILL BE ONLY ONE MONETARY SYSTEM

February 29, 1960, MELCHIZEDEK said: "ONLY ONE KIND OF MONEY WILL BE USED WHEN THERE WILL BE BUT ONE WORLD GOVERNMENT."

September 28, 1960, ROCK OF AGES said: "Money will not have the power it once had.

"IN TIME TO COME, ALL OF THE DIFFERENT MONETARY SYSTEMS USED BY THE VARIOUS NATIONS WILL BE CHANGED."

November 24, 1961, A Great Being named IRIAH, said: "All Dictator Leaders of Governments will be removed.

"All of the Monetary Systems will be changed in every nation on the earth. In about ten years these changes will be in the gradual process of coming into operation."

July 27, 1962, SETHANTES said: "The time is at hand when all Monetary Systems will be in the throes of change. Everything is in a turmoil . . . in confusion."

August 23, 1962, NOSTRADAMUS said: "You have been told there will be a NEW MONETARY SYSTEM. It will be a long drawn-out affair, because it is a 'Principle' that will operate in every Nation on the earth. The money will be of the same denomination in all Nations, but it will not have the power that it always has had. Money has been the cause of most of the wickedness on the earth. People will do drastic things for money. They rob and murder because money has that power over an individual person. In this New Age, money will be dispensed with."

(In the year 1973, the TRI-LATERAL ORGANIZATION was formed to create a NEW ECONOMIC ORDER.)

On March 12, 1979, a News Headline stated:

("NEW EUROPE MONETARY SYSTEM STARTS TOMORROW." It is designed to stabilize currency fluctuations within the eight-member "Common Market." This was announced at a two-day Summit Meeting. The eight countries involved are: West Germany France, Italy, Belgium, Luxemberg, Holland, Ireland and Denmark.)

ONE WORLD GOVERNMENT . . .
ONE WORLD UNITED STATES

February 29, 1960, MELCHIZEDEK said: "IN THE FUTURE, DURING THIS NEW AGE, THERE WILL BE BUT ONE WORLD GOVERNMENT ON THE EARTH. IT WILL BE ONE OF RIGHTEOUSNESS!"

September 28, 1960, ROCK OF AGES said: "The Nations of all the earth are going to make this entire earth world a 'ONE NATION.' "

November 10, 1961, Sara psychically saw an IMMORTAL BEING in the room, and it is his first contact at our center. She described him: "He wears a symbolical headdress that looks like a building and there is a light inside of it. I also see the whole sky filled with those BEINGS wearing that same kind of a 'building headdress' and the lights in their buildings 'blink on and off.' "

Then the GREAT BEING in the room said: "Construction is underlying destruction. That is why we are here and in the atmosphere. The earth itself is undergoing a great change, and the constructive changes now taking place are far more potent than the destructive changes.

"There will come a day in the future when the President of the United States will sit down with the Leader of the Soviet Union of Russia. Together they will work out a plan that will change events on the earth.

"The time is coming, and it is not too many years hence, when you will have 'A WORLD UNITED STATES.' Then the whole earth world will be operated in a righteous way, such as the United States of America will then be demonstrating at that time.

IN TIME TO COME, THERE WILL BE AN ORGANIZA-

TION OF THE EUROPEAN NATIONS, KNOWN AS THE EUROPEAN UNITED STATES."

April 19, 1962, A Great Master named STAR OF BETHLEHEM, said: "THE UNITED STATES OF EUROPE IS BEGINNING TO FUNCTION. THE NATO TREATY ORGANIZATION IS THE 'SPARK' OF THAT WHICH, IN TIME TO COME, WILL BE KNOWN AS THE UNITED STATES OF EUROPE.

"THE NATO COUNTRIES WILL ALIGN THEMSELVES WITH THE UNITED STATES OF AMERICA."

MICHAEL'S NEW AGE PREDICTIONS

January 15, 1964, MICHAEL, A GREAT MASTER, said: "All things throughout the earth world are gradually changing. The constructive vibrations are underlying and operating underneath the surface. Those constructive changes are pushing the destructive things forward into the open, so that all earth people can see the evil things that are transpiring. It won't be too long until you will see the many new future changes that you have been told about. They will be exciting and wonderful. Those changes will seem very strange to all mankind on the earth.

"When the year 2,000 arrives, then you will see that the sky will have become tinted with lavender, and this also applies to the atmosphere. By that time the atmosphere will have a feeling of affection . . . and of spiritual love . . . that will touch the hearts of all mortal earth people.

"The trees will become great tall trees, such as have never appeared on the earth before that time. The trees will take on different hues and colors, such as blue and pink. The shrubs will not be green as they now are. They will have different hues and colors. This also applies to the landscape which will become very colorful.

"The physical bodies of mortals will have become changed, perhaps not so much in appearance, but in texture and quality. Also, the Immortal Spirit Beings who will mingle with the mortals on the earth, will undergo changes.

The spirit world and the earth world will become blended together as 'ONE AND THE SAME PLACE.'

"The earth world will have become entirely different, and all things will have been made NEW. Then, earth people will experience Peace and Harmony, and Great Spiritual Love for one another, when the year 2,000 arrives.

"Blessed are those people who have attained the 'Light of Understanding' and who walk in the Light.

"Little do the people on the earth realize WHAT is ahead of them. SO WATCH . . . and . . . WAIT!"

The DIVINE ONE, A GREAT MASTER, said: "My eyes reach out into the Universe, and I behold what the CREATOR has in store for all mankind." (1-11-62).

THE SEVEN MYSTICAL CHRISTS

On November 13, 1959, THE MASTER JESUS, spoke through Sara, and gave us the following message:

"The ALMIGHTY CREATOR creates SPECIAL HUMAN BEINGS to perform specific work in HIS Vast Cosmic Schemes of Creations. Those BEINGS are limited in number, and they are set apart from all the rest of HIS countless created Human Beings. HE endows them with all of HIS own Inherent Spiritual Attributes. Those Attributes include all of HIS Mighty Creative Powers, Infinite Wisdom, and Divine Spiritual Love . . . for all mankind. They receive those Mystical Qualities at the time of their special creation.

"There are but SEVEN MYSTICAL CHRISTS who have been created by the CREATOR, to perform their mystical work in HIS countless millions of worlds existing in HIS Vast Cosmic Schemes of Creations.

Those SEVEN CHRISTS take their places on seven different plateaus of solar systems, and there is a vast empty space in between each of those plateaus. Over each plateau the CREATOR has placed 'One Christ' who rules all those millions of worlds comprising one plateau.

"After the year 2,000 is completed, then the Christ who will be ruling the earth, will demonstrate his powers, which will operate on the earth and in the earth's heavens. His powers will bring into full operation the NEW HEAVEN and the NEW EARTH.

"No person on the earth has ever seen a 'Christ' who emanates his Glorious Golden Light. Perhaps, some people have psychically seen a Great Immortal Master who they believed was a Christ. I, JESUS, function through several different Personalities when I come to the earth. But I have never appeared as a Christ.

"I AM A CHRIST and at some future time I will function through my Christ Personality. Then I will be known as the CHRIST who rules over the solar system containing the earth.

"A CHRIST IS AN ORIGINAL ARCHETYPE MODEL OF A HUMAN BEING, WHO DISPLAYS HIS CHRIST CONSCIOUSNESS. HE OPERATES IN THE SAME WAY IN WHICH THE CREATOR OPERATES.

"The CHRIST, as a HUMAN BEING, or as a POWER, will not start to manifest and to demonstrate until after the year 2,000. That is the time when the NEW EARTH and ITS NEW HEAVEN will begin to operate as 'One Condition.' That also is the time when the CHRIST will take his Rule and his Reign. He will demonstrate an entirely different way of life for earth people. Then, at that time, all mankind on the earth and in its heavens, will be different people. Because the CHRIST POWER will enter into every Human Being, be he a mortal, or an immortal. The CHRIST POWER will demonstrate and take over each person.

"THE CHRIST POWER IS THE CREATIVE POWER!

"No person on the earth knows the 'SECRET OF CHRISTHOOD.' I operate in the CHRIST CONSCIOUSNESS. In the future, the earth will become a world in which the Christ Consciousness will demonstrate and operate.

"AFTER THE YEAR 2,000, THAT IS THE TIME WHEN THE 'GREAT EVENT' WILL TAKE PLACE. AS 'THE CHRIST" . . . I WILL APPEAR IN THE SKY AND ALL PEOPLE WILL SEE ME. THAT IS WHEN I WILL START MY RULE AS 'CHRIST.' I WILL RULE FOR 1,000 YEARS, AND BE KNOWN AS THE CHRIST OF THE SOLAR SYSTEM CONTAINING THE EARTH!

"By that time earth people will have attained the Christ Consciousness. Then, the CHRIST LIFE will operate and manifest what the CREATOR has planned for this solar system.

"The CHRIST LIFE is something that operates entirely in spirituality. It brings you into the CREATIVE POWER, and into the full realization that you are indeed a Creator in your own right.

"At the end of 2,000 years, then the earth people now known as earth mortals, will become known as Christ Beings. All the necessary changes will have been made in their bodies. They will have the attributes, creative powers, and all the development they will need to function as Christ Beings.

"The SEVEN CHRISTS are here tonight, and this is something that has never before occurred in all of creation. Tonight, your minds have been lifted into higher realms of consciousness.

"THE PICTURE IS UNFOLDING . . . THE WHOLE PLAN OF THE CREATOR IS BEING UNFOLDED RIGHT HERE IN YOUR MIDST! WATCH . . . and . . . WAIT."

CONCLUSION

May 30, 1958, Sethantes said: "The greatest events in history makes no more noise than the rising of the Morning Star. The greatest events are the silent ones."

August 22, 1958, The God of All Pyramid Worlds, made this statement:

"THE GREATEST EVENT THAT HAS HAPPENED ON THE EARTH WORLD SINCE THE BIRTH OF THE MASTER JESUS, HAS TAKEN PLACE RIGHT HERE IN THIS LITTLE CENTER, IN LONG BEACH, CALIFORNIA."

* * *

It is a basic fact that we live in an enormously, vast, complex universe, that is designed and governed through precise, orderly universal laws. We marvel, as we contemplate upon the awesome vastness of the limitless extent of space, containing its numerous universes, suns, moons and countless billions of stars. All, systematically whirling and moving together in harmonious rhythm, like figures in a mysterious dance. Therefore, logical reasoning leads to the belief of the existence of "A MYSTERIOUS, SUPREMELY INTELLIGENT DESIGNER . . . AN OMNIPOTENT ALMIGHTY CREATOR."

All Adoration, Praise and Glory to THEE, O BOUNDLESS ALMIGHTY CREATOR!

Rev. Sarah Ferguson (right, in white), was Ordained by Rev. Lola Reddig, June 25, 1950, in Long Beach, California.

When this photo was developed, the Spirit photo, upper right, appeared. Later, June 17, 1957, Sarah's Spirit Teacher "Adontus" said: "That is my picture. I have been in the spirit world for more than 50,000 years." Sarah passed into spirit life November 3, 1972, age 94 years.

Sarah Ferguson and Dwight D. Eisenhower.

They were first cousins. This photo Sarah had explained to me was taken by a news photographer at the Long Beach, California airport. From Sarah's writing on the photo, it appears to state that photo was taken in 1946.

Sarah gave it to me, Laura Horst in year 1960. Believe Sarah was about age 48 at that time. Sarah died at age 94 years. She was born October 9, 1878.

—by Laura Horst

SARAH (SARA) FERGUSON'S LIFE STORY

Sarah Ferguson's grandfather, Jacob Eisenhower, and his wife Rebecca, in 1854 built their nine-room brick home on a one-hundred acre farm near Elizabethville, Pa. He was a farmer, and also a Pastor of a religious sect, known as the "Brethren in Christ," who were sturdy farm folks. He preached "Brotherly Love," and his home was a meeting place for the sect's religious worship.

From the union of Jacob and Rebecca, fourteen children were born, and among the seven who became adults were David and Susan. Jacob was kind, but he used strict discipline in ruling his family, and all conformed to his religious teaching.

A large colony of the Brethren Group migrated to Kansas, and Jacob moved his entire family, excepting Susan, to Abilene, Kansas, in the year of 1878, and they settled on a 160-acre farm. He also acquired a farm of 160-acres for his children.

Jacob's son David married Ida Stover, and their son Dwight David Eisenhower, became the 34th President of the United States.

Jacob's daughter Susan married Aaron Wetzel, and Sarah is one of their ten children. Sarah (Sadie) Wetzel was born October 9, 1878, in Elizabethville, Pa. When she was four years old her family also moved to Abilene, Kansas, but her father disliked farming, so they moved to Kansas for only a few years, and then they returned to Pennsylvania.

Sarah Wetzel married Joseph Vogel, and two sons, Harry and Carl, were born to them. Years later, after the passing of Joseph, Sarah married Raymond Ferguson, on December

24, 1910, in York, Pa. They, and their young son, moved to Long Beach, California, in the year of 1924. Sarah was a dressmaker, and for many years she operated her modest little dress shop in Belmont Shore, in Long Beach.

There were several Spiritualist Churches in Long Beach, and in the year 1925 Sarah became interested in the psychic phenomena demonstrated by the mediums.

It is interesting to know how Sarah developed her mediumship. She had, from early childhood, studied the Holy Bible, as her grandfather was a very devout Protestant Minister, and he saw to it that his children and grandchildren attended his church. He did not, however, teach that "Spirit Communication" was possible. That is; that the Immortal Spirits, the so-called "dead" can communicate with earth-mortals, if they have the proper channel which they can use.

In March, 1932, Sarah met Elizabeth Gainor, a Trumpet Medium, at a circle-seance meeting. At that time, Elizabeth's SPIRIT-TEACHER, DR. BOUNDS, introduced Sarah to her MASTER SPIRIT-TEACHER, "ADONTUS." Elizabeth lived in Santa Barbara, but occasionally she came to Long Beach, and held private trumpet seances, which Sarah attended. At different times she would stay overnight with Sarah in her home, and they held private seances.

To enhance her spiritual unfoldment, Sarah would sit alone and meditate each morning, from 5-a.m. to 6-a.m., and also meditate during one hour each evening, sitting in silence, in a small dark clothes-closet, with eyes closed, meditating and asking for spiritual enlightenment. During those meditation periods she discovered she had psychic sight. With her eyes closed, she could plainly see into the spirit-side of life, and psychically see the "visions" that her Spirit-Teachers would show to her. For many years she faithfully meditated each day.

During a group-circle meeting in 1933, a GREAT MASTER TEACHER spoke through Sarah. This was the first time she was used as a "channel," and at that time she received her Gift of Mental-Mediumship. Thereafter, her Spirit-Teachers used her many times, as a communicating channel, speaking through her vocal-cords. She asked them to give her greater inspirational light, as she desired to learn more about the deeper things of life.

Raymond Ferguson's brother Earl came to live with Ray and Sarah, in the year of 1934. Earl became interested in psychic phenomena, and he attended Mrs. Gainor's trumpet seances whenever she came to Sarah's home. The Spirit-Teachers spoke to them through the Trumpet, and sometimes they used Independent Voice, or Direct Voice. "PO," the GREAT ONE, spoken of in Oahspe, announced that he was Earl's Master-Teacher. (Mrs. Gainor made the transition in 1944).

During the years 1934 to 1947, Earl would sit with Sarah in her little sewing room, in her home, and they would hold a seance. He was like a "battery" for her, as his presence brought added vital odic-power. Using Sarah as a channel, numerous GREAT SPIRIT-TEACHERS told them many very astounding things, and many "visions" were shown to Sarah, and she would describe the visions to Earl. Many times, after a session, Earl would write in a little book the names of the TEACHERS who spoke, and also write whatever information he could remember. Sarah and Earl did not fully understand what was taking place, but they always looked forward to each session. Sarah saved some of Earl's writings, and gave them to Laura Horst, who has them in her files.

From the time of Earl's transition, in 1947, until about June of the year 1956, Sarah seldom held a session, as she had no other person who was interested enough to attend regular sessions. And her husband, during that period of time, was not interested in psychic phenomena. However, during all of those intervening years, Sarah faithfully meditated, and communed with her Beloved Spirit-Teachers. During that time, her psychic development grew by leaps and bounds. And "PO" told her that he now was with her as a TEACHER, and as a co-worker with "ADONTUS."

During the year of 1957, Sarah's husband Ray, became interested in the sessions that Sarah conducted in their home, and thereafter he eagerly attended weekly sessions. Among his Master Spirit-Teachers were the GREAT OSIRIS and APOLLO, who are spoken about in Oahspe. (Ray made his sudden transition on June 13, 1964).

Sarah was ordained as a minister on June 25, 1950, by Reverend Lola Reddig, the Paster of The Temple of Christian Philosophy, in Long Beach. After the Ordination Ceremony, Lola and Sarah were photographed, and when the photo was developed, very plainly there appeared upon the photo, near Sarah, the head and face of a Spirit-Being, an aged man. (Later, ADONTUS told Sarah it was his picture).

Reverend Lola Reddig was a gifted medium, having the Gifts of Trumpet, Direct Voice, and Materialization. She held a materialization seance once each month, which Sarah attended. Starting in June of 1951, and continuing each month for seven months, THE SEVEN GOLDEN AGE MASTERS materialized. That is; each month, One Master would appear, until all seven had appeared. Each Master, as he appeared, would stand before Sarah, and give his name, and then he would give an important message to the group. The Masters told WHY they were appearing at this time; saying: "Thousands of years ago, there was a GOLDEN AGE on the earth, and we are now entering another GOLDEN AGE . . . THAT IS WHY WE ARE HERE!" Sarah did not understand WHY they came to her, and she could not record their messages.

These are the names of the SEVEN GOLDEN AGE MASTERS: Apollo, Ashbina, Fragapatti, Kaspar, Matheno, Megsta', and Philo.

During June of 1956, Catherine McGuire started to attend regular weekly sessions with Sarah, in the Ferguson home. Among many GREAT ONES who spoke through Sarah, was the GREAT MASTER "SUE," spoken of in Oahspe, and he told Catherine he was her Spirit-Teacher. During their private sessions, Catherine and Sarah received much comprehensive important information. And for many years, they faithfully continued their sessions.

The Spirit-Teachers have highly magnetized Sarah's body, and during many years, they have used the psychic-power drawn from her body to perform many phases of their work, both on earth and in the spirit-world. They have used her power to help them remove discarnate Spirit-Entities from earth-mortals. They also use the power which they draw from other persons who attend a session. They use all the psychic power available from mortals; for healing work, and for releasing earth-bound immortal Beings, the so-called "dead," and then they take them into a higher progression of life in the spirit-world. The SPIRIT-TEACHERS need mortals to help them with their GREAT WORK.

Sarah and Laura Horst became acquainted in December, 1956, and Laura's story briefly tells about the many private sessions she attended in Sarah's home. Sarah's Spirit-Teacher "PO," and Laura's Spirit-Teacher "SETHANTES," the GREAT BEINGS spoken of in Oahspe, and OTHER GREAT ONES, gave them very important information. Those Horst-Ferguson sessions were held at an average of twice each week, and covered a period of time from January 1957, through the year of 1964, and only a very few sessions were held thereafter. All of those very important sessions were tape-recorded and transcribed, and all of the original transcripts, together with numerous tapes, are in Laura's files.

SARAH FERGUSON TRULY IS A VERY REMARKABLE PERSON, AND SHE WILL CELEBRATE HER 92nd BIRTHDAY ON OCTOBER 9, 1970.

This story written by Laura Horst, September 23, 1970.

Approved and signed by Sarah Ferguson.

ADDENDUM to Sarah's LIFE STORY
from items in her files

October, 1933, Sarah's Master Teacher "ADONTUS" said to her:

"The clouds above your head are as vapor, and of life substance were you created. The ALMIGHTY CREATOR created you in Goodness and in Love. HE gave you, Wisdom, Love and Power, to be used for the good of your fellowmen. You are to be scoured and polished, like a diamond in the rough is cut and polished, and so must you be treated likewise. And this applies to all mortal Human Beings, who are in their embryonic stage of life. A seed planted today, does not become a tree tomorrow. It requires a slow steady growth to become a strong tree and to bear good fruit. Each person is tried and tested, as each needs many rough experiences, in which to strengthen the spiritual qualifications of his Own Soul.

"The magnetic electrical force that is the vibration emanating from your body forms your 'aura' surrounding you, and it reveals your spiritual unfoldment. Every Human Being who seeks greater knowledge of LIFE, links-in with the vibration of other Human Beings, who have become raised higher in their spiritual unfoldment. The Higher Raised Ones, always helping the Lower Ones. And so it goes on and on with all people, through All the Ages of Eternity.

"The 'MIND' is the 'SPARK OF DIVINITY' that is part of the CREATOR. And 'THOUGHTS' are the substance flowing through the 'MIND.' And the 'THOUGHTS' tie-in with the 'WILLPOWER.' These things stem from the CREATOR. Your own thread of your Real Life . . . is YOU, YOURSELF . . . YOUR OWN 'I AM' . . . THAT SPARK OF DIVINITY, THAT COMES FROM THE ALMIGHTY CREATOR, WHO CREATED YOU."

She passed into spirit life, November 3, 1972, at age 94 years.

THE COSMIC DRAMA OF LIFE

O, Almighty Creator — Thou Great Central Light
of ALL LIGHT!
Thou Supreme Creator of all creations, of all
universes; and all things they contain;
Thou Loving Almighty Father; Creator of all Gods;
and Lords and all Human Beings and all living
things!
The Heavens declare Thy Glory; and I see Thy
Miraculous Wonders everywhere!
Beloved Father — guide and teach me Thy Holy
Wisdom!
O, Almighty Creator — Thou art within me and
I am within Thee.
Thou hast given me my own Eternal Soul — a
Luminous Spark from Thy Eternal Radiant Light,
illuming my Soul and forever joining my
Soul to Thee.
I feel our sacred Oneness — and deep within my
own Soul, I hear Thy Loving Whispers from
Eternity.
With loving reverence, praise and rejoicing
within my Soul, I watch Thy Cosmic Drama of
Life — ever moving and changing as a vast,
endless dream.
Glory to Thee, Beloved Creator, forever and
forever!

—by Laura Horst
Year 1963

Laura (Lara) Horst
Age 30

Yr 1930